Introducing
Robotics with
Lego® MindStorms™

Other titles of interest

BP902   More Advanced Robotics with Lego MindStorms

*   *   *   *   *

BP450   How to Expand and Upgrade Your PC
BP467   How to Interface PCs
BP470   Linux for Windows Users
BP479   How to Build Your Own PC
BP484   Easy PC Troubleshooting

*   *   *   *   *

BP394   Programming Visual BASIC for Windows
BP444   Windows 98 Explained

*   *   *   *   *

BP394   An Introduction to PIC Microcontrollers
BP444   Practical PIC Microcontroller Projects

Introducing Robotics with Lego® MindStorms ™

Robert Penfold

Bernard Babani (publishing) Ltd
The Grampians
Shepherds Bush Road
London W6 7NF
England

Please note

Although every care has been taken with the production of this book to ensure that any projects, designs, modifications, and/or programs, etc., contained herewith, operate in a correct and safe manner and also that any components specified are normally available in Great Britain, the Publisher and Author do not accept responsibility in any way for the failure (including fault in design) of any projects, design, modification, or program to work correctly or to cause damage to any equipment that it may be connected to or used in conjunction with, or in respect of any other damage or injury that may be caused, nor do the Publishers accept responsibility in any way for the failure to obtain specified components.

Notice is also given that if any equipment that is still under warranty is modified in any way or used or connected with home-built equipment then that warranty may be void.

© 2000 BERNARD BABANI (publishing) LTD

First Published - August 2000

British Library Cataloguing in Publication Data
A catalogue record for this book is available from the British Library

ISBN   0 85934 901 2

Cover Design by Gregor Arthur

Printed and bound in Great Britain by Bath Press

Preface

On the face of it the LEGO® MindStorms™ Robotics Invention kit is not that much different to many other kits that have been produced over the years. You snap various bit and pieces together to build model cars, tractors, cranes, or practically any other piece of machinery. The electric motors supplied with the kit make things work, but this is again nothing new. What makes the MindStorms kit different is the RCX unit, which is an outsize Lego brick that contains a microcontroller. Not only do the finished models work, but they can also "think" for themselves. Rather than just going round in circles, finished models can perform pre-programmed routines, react to input from their sensors, and (or) they can be controlled manually from a PC via an infrared link. Two robots can even "talk" to each other via an infrared link. We have not yet reached the stage where you can partially build a robot and then get it to complete itself, but the finished robots can be programmed to perform quite complex tasks. These include games and "party tricks".

Programming a robot is potentially a very difficult task, but the MindStorms kit is supplied with a programming language that greatly simplifies this task. Conventional programming starts with a flowchart, and then the code to implement each part of the chart is written. With RCX code the flowchart is produced on a PC, and it is automatically turned into code that can be downloaded to the RCX unit. This simple method makes it easy for non-programmers to get started, but it can be a bit limiting for experienced programmers. The MindStorms system is also supplied with an ActiveX control that enables more sophisticated programming using a conventional Windows programming language such as Visual BASIC. Both methods of programming are covered in this book, with plenty of practical examples that demonstrate the control of the motors, reading light and touch sensors, using variables, etc. Taking direct control of robots using Visual BASIC and the infrared link is also included.

The mechanical side of things is not overlooked. Although the snap together system used for Lego makes things as simple as possible, with 717 pieces in the kit this is a potentially baffling aspect of things for beginners. Chapter one describes methods of building strong structures using the plates, beams, and the other simple components. It also covers the various types of gearing available, using pulleys and drive bands, and uses for some of the more obscure components in the kit. Building the example robots is straightforward due to the inclusion of detailed descriptions backed up by numerous stage-by-stage photographs. It is assumed that the reader has gone through the multimedia introduction to Lego Mindstorms provided with the kit, and can handle basic tasks such as downloading RCX code to a robot. No skills beyond this are needed to follow the examples provided in this book. None of the example robots require any parts that are not included in the standard Robotics Invention Kit.

Whether you are interested in the MindStorms kit as a toy for having fun, or as an aid to learning about robots, this book provides a solid introduction to building the robots and programming them.

Robert Penfold

TRADEMARKS

Lego® and MindStorms™ are registered trademarks or trademarks of the Lego group of companies. All other product and brand names may be registered and legally protected trademarks of the companies that manufacture those products. There is no intent to use any trademarked name generically or to infringe that trademark in any way, and the names are only used in an editorial or literary context. Readers are advised to fully investigate ownership of a trademark before using it for any purposes.

To aid readability of this book, the trademark symbols ® and ™ are used initially, but not at every occurrence of the trademark name. This does not imply in any way that the name may not be a trademark or registered trademark.

Visual BASIC and VBA are registered trademarks of Microsoft Corporation

Delphi is a trademark of Borland International Corporation

Pentium is a registered trademark of Intel Corporation

UNOFFICIAL GUIDE

Neither the Author or Publishers of this book are connected or supported in any way by any Lego group company, and should not be confused with any publication that may be produced by a Lego group company.

Contents

1 Mechanics 1

2 Using Spirit.OCX 23

3

Lightbots 59

5

Direct control 149

Mechanics

Beams and plates

Designing structures using Lego is not exactly difficult, but producing strong structures that will not keep falling apart is slightly less straightforward. Another problem for beginners is the number of parts in the Robotics Invention System. I have not counted to check, but the claimed 717 pieces in the basic kit seems to be eminently plausible, and the total will soon exceed 1000 if you start adding some expansion kits. With such a large number of items it inevitably takes a while to fathom out what everything does, and how to combine the components effectively. In this chapter we will look at the ways some simple but strong structures can be produced. The all-important subject of motors and drive systems is also covered in some detail.

Beams and plates form the basis of Lego structures, which almost invariably consist of beams held together by plates. The beams are mostly black in colour, and are one unit wide and up to 16 units long. When dealing with Lego beams and plates it is normal to specify their size in terms of the number of units (lumps/holes) rather than a size in millimetres. This is simply because the units method makes it much quicker and easier to locate and identify the required pieces, since no measurements have to be made. The beams come in two forms, which are the plain variety and those that have holes going right through them. Only the shortest beams lack the holes. The holes through the beams serve two purposes, one of which is to take axles. The other is to take small grey and black plastic pieces that can be used to fix beams to each other or to certain other Lego components. I do not know if there is a correct name for these, but for want of a better term they will be referred to as pegs or "rivets" in this publication.

The plates are mostly made from grey plastic, but there are a few green and yellow ones in the basic kit. They range in size from 2 x 1 to 10 x 6. Some have holes to take axles, etc., and some are plain. Lego is very versatile, and robots built using the Robotics Invention System will obviously vary considerably in size and shape. However, all robots that use the RCX unit must have a platform to take this sizeable Lego "brick", and this platform is normally constructed from beams and plates. In most cases this platform is also the chassis of the robot, and it therefore takes motors and wheels in addition to the RCX unit.

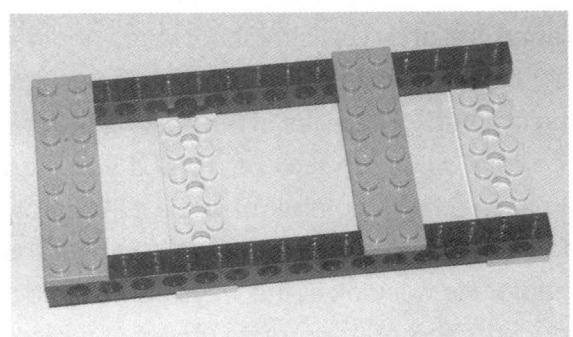

Fig.1.1 A very simple chassis

Fig.1.2 A simple but strong chassis

Chassis building

At its most basic a chassis can consist of two beams joined by two plates on top of the beams (Figure 1.1). The plates can provide a suitable mounting for the RCX unit. The problem with a chassis as simple as this is that only a minimal amount of force is needed in order to break the beams away from the plates. Greater strength can be obtained using further plates to join the beams, with perhaps two more on the underside of the assembly. Further strength can be added by using two beams on each side of the chassis, with "rivets" used to join each pair of beams. Each plate is then joined to the chassis by a 2 x 2 area instead of a 2 x 1 area, which gives greatly improved strength. It is not essential to use two identical beams in each side-piece, and short beams can be added to a longer type at the places the plates are fitted. This gives an assembly of the type shown in Figure 1.2. The increase in strength provided by this sort of thing is probably much greater than you would expect. Try it for yourself, and you will soon see what I mean.

Long and short of it

The longest beams in the kit are 16 units long, but it will sometimes be necessary to have a chassis that is somewhat longer than this. There are two main ways of joining beams together end to end to effectively form one long beam. The simplest method is to use plates to join them together, as in the structure shown in Figure 1.3. Provided plates are used both above and below the beams and the plates are quite long, this method works reasonably well.

The method preferred by most is to use another beam plus at least four pegs, as in Figure 1.4. An advantage of this system is that it gives the doubling up of the beams in each

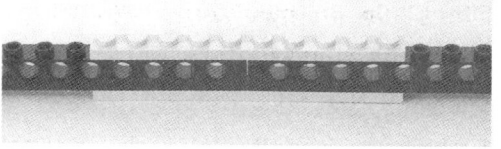

Fig.1.3 Joining beams using plates

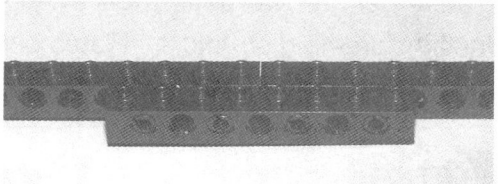

Fig.1.4 Joining two beams using pegs plus another beam

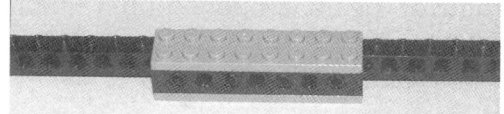

Fig.1.5 Using both methods to join two beams

Fig.1.6 Using cross-beams to strengthen a chassis

side-piece, as described previously. This enables large but strong chassis to be produced. A third option is to combine the other two methods, as shown in Figure 1.5. This "belt and braces" approach gives the ultimate in strength.

Anything that makes a structure more solid should help to give it increased strength. For example, when joining the two sides of a chassis together beams can be used in addition to the plates (Figure 1.6). It is advisable to use some plates that are two units wide when joining side-pieces, since this gives a much more rigid structure. Plates one unit wide permit the chassis to skew out of shape.

Vertical hold

It is often necessary to build up a chassis vertically, perhaps in order to move an axle lower on the chassis or a motor higher up, so that a gap for the required gears is opened up. Simply adding one beam on top of another gives good strength. In fact there is usually no problem in piling up several beams one on top of another. There is a slight problem if some additional strengthening is required, or if the structure requires beams to be fitted vertically. The obvious method of pegging a vertical beam or beams to horizontal types does not work. A vertical block unit is slightly larger than a horizontal unit. When beams are fitted one on top of the other the vertical spacing between holes is therefore greater than the horizontal spacing. It is possible to get around this by adding in one or two plates to adjust the hole spacing so that everything fits together properly. You can then peg either beams or the grey and black girder-like pieces to the structure in a vertical position (Figure 1.7).

Fig.1.7 Using vertical beams to strengthen a structure

Mounting motors

There are several types of motors in the Lego system, but only the geared type are normally used with the MindStorms kits. Both the motors supplied with the Robotics Invention System are the geared variety. The name is derived from the fact that this type of motor has integral step-down gearing. Mounting motors on your robots is not quite as easy as it might at first appear, due to the slight bulge on what is normally the underside of the motor (Figure 1.8). If you make a nice flat platform for the motor to stand on, it will not attach to it properly. In fact the slight bulge will prevent it from fitting onto the platform at all.

Fig.1.8 The bulge underneath a motor

It is essential to have the motors firmly fixed in place, otherwise they will simply pop out of position as soon as they are started. The underside of each motor has what is effectively a main 4 x 2 fixing pad, plus two 2 x 1 pads. Ideally this entire area should be used to fix each motor to the chassis, but this will not always be practical.

Fig.1.9 A mounting pad for a motor

To fix a motor onto a platform it is merely necessary to add some plates to form a mounting pad, as in Figure 1.9. This only raises the motor by about three millimetres, but this is enough to keep the motor's bulge clear of the main platform, so that the motor fits into place properly (Figure 1.10).

Fig.1.10 A motor fitted onto the mounting pad

In amongst the small items in the kit you will find 2 x 1 plates that have small flanges protruding from one of the longer sides. The flanges fit into the slits at the sides of the motors, and they can provide additional fixing for the motors. A very simple way of using them is to use two of these plates with a 2 x 1 plate fitted underneath. They can then provide more secure mounting, as shown in Figure 1.11.

If a 2 x 1 beam and another flanged 2 x 1 plate is added to each side, you have an assembly that will fit into all four slots of a motor. With the aid of three plates it is then possible to fix the motor between the two side-pieces of a chassis, as in Figure 1.12. When doing this type of thing make sure that the black 2 x 2 area on top of the motor does not become obscured. This is where the connections to the motor are made. Note that the flanged plates should not be used as the sole method of mounting a motor. They slide into the slots in the motor, and without some additional fixing they will slide out again at the earliest opportunity.

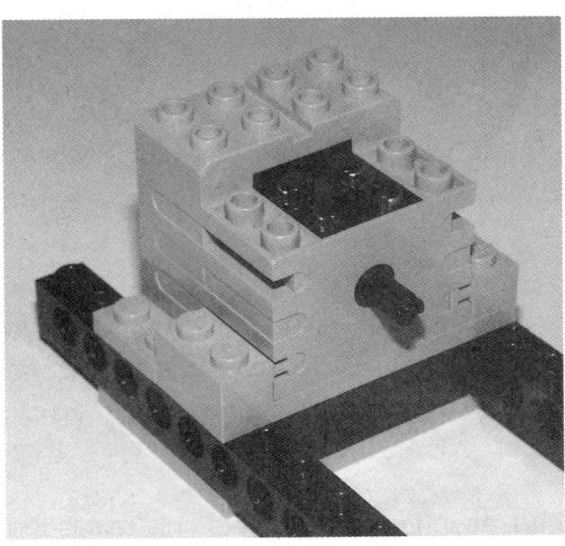

Fig.1.11 Using the slots in the motor for more secure mounting

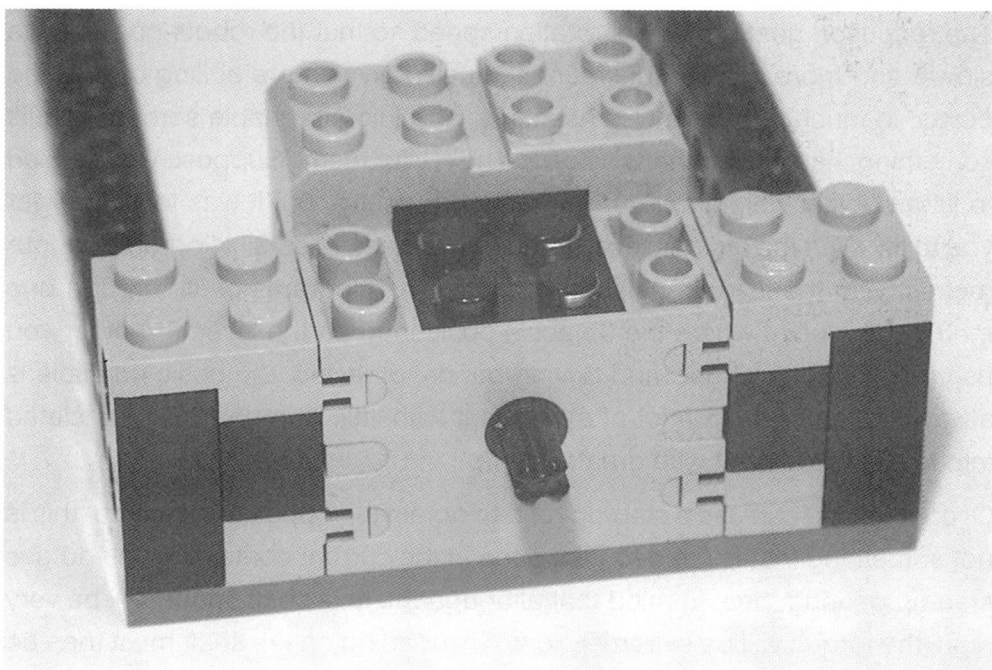

Fig.1.12 A very secure method of mounting a motor

Why use gears

Gears are mainly used where the raw speed of the motor is inappropriate. By using a 2 to 1 reduction gear for example, the speed of the motor is effectively halved. Although the rotation speed is reduced, if we ignore any losses through the gears, there is no reduction is the available power. The reduction in speed is matched by an increase in torque. As already pointed out, the motors supplied with the MindStorms kits have built-in reduction gearing. This is done because miniature electric motors operate at high speeds but have little torque. When applied to most robots this equates to things happening too quickly if the motor is loaded lightly. With the loading increased to a moderate level it is likely that the motor would stall and the robot would fail to work at all. With a robot arm for instance, the arm would tend to swing into position very rapidly, probably overshooting its intended stopping point by a considerable margin in many cases. Having grasped an object to be moved it would then have insufficient torque to actually move it.

The reduction gearing lowers rotation speed so that the robots operate in a slower and more controllable manner. It also gives more pulling or pushing power, in much the same way as a lever and fulcrum enable someone to lift something that would otherwise be too heavy for them. Suppose you needed to lift an object weighing 100 kilograms one metre, but it was too heavy for you to lift. With the aid of a lever you could instead raise 25 kilograms by four metres, with the lever translating this into 100 kilograms of lift through one metre at the point where the object is positioned on the lever. Whether you use gears, pulley wheels and driving bands, or levers, the basic principle is always the same, with a lot of movement with little oomph being translated into lesser movement with greater oomph.

Of course, you can use a step-up ratio to obtain the opposite effect, but this is not something that is needed very often in the current context. If you do use step-up gearing, bear in mind that although the final shaft speed will be very high, the torque will be extremely low. The loading on the shaft must then be kept to a minimum, or rather than rotating at high speed it will simply fail to rotate at all.

Gears are not only used when a step-up or step-down ratio is required. Particularly when using the smaller wheels, direct drive from the motors to the wheels may be perfectly all right in theory, but in practice it can be difficult to find a satisfactory form of construction. There is no point in having the wheels fitted on the shafts of the motors if the wheels will not reach the ground! Sometimes gears have to be used simply to get the power to where you need it. Of course, in such cases a gearing ratio of 1 to 1 can be used..

Types of gearing

The simplest form of gearing has two ordinary gearwheels, as in the example of Figure 1.13. The ratio of the gearing depends on the number of teeth on the gearwheels, and the Lego gears have 8, 16, 24, or 40 teeth (Figure 1.14). Using a single pair of gears this gives a maximum step-up and step-down ratio of 5 to 1 with the 40 and 8 tooth gears (40 divided by 8= 5). Having the larger gearwheel on the drive shaft gives a step-up ratio, and having the smaller gear on the drive shaft produces a step-down ratio.

Of course, if you simply wish to transfer power from one shaft to another with no step-up or step-down, use gearwheels of the same size to obtain a 1 to 1 ratio. When using this type of gearing do not forget that the driven shaft moves in the opposite direction to the drive shaft. This is not necessarily of any great consequence in the current context, since the motors can be made to turn one way or the other depending on how they are connected to the RCX unit.

Fig.1.13 Simple reduction gearing

You are not limited to having a simple transfer of power direct from one shaft to another. Using multiple gearing, or compound gearing as it is often termed, it is possible to have one shaft drive several others, or to have power transferred via two or more sets of gears. Where one motor

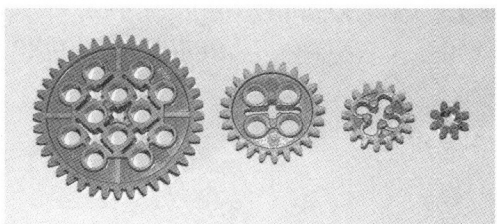

Fig.1.14 The four standard Lego gearwheels

is used to power several shafts you must always bear in mind that the power available from a single motor is limited. Something that looks fine "on paper" might not actually work in practice if the motor is loaded excessively.

There are two main reasons for transferring power through more than one set of gears. One is simply to bridge a larger gap than a single set of gears can accommodate. Using two 40-tooth gears a shift of five units is obtained, but this might not be enough, or there might not be sufficient space for the large gearwheels. Two smaller sets of gears would then be a more practical proposition. Providing large shifts using numerous gears is probably not the right way of handling things. In most cases it would be better to use a different form of gearing, switch to pulleys and drive bands, or redesign the robot to bring the motor closer to the action. Remember that there are small power losses through each set of gears due to friction. This is unlikely to be of significance with two or three sets of gears, but could produce a noticeable loss of power if there are sets of gears here, there, and everywhere.

Fig.1.15 The initial stage of the compound gearing example

Fig.1.16 The final stage of the compound gearing

More ratios

Using a single set of gears it is only possible to obtain a limited number of drive ratios. By utilizing two or three sets of gears the number of available ratios becomes vast. Suppose you need a step-down ratio of 15 to 1. As already pointed out, the maximum ratio for a single set of gears is 5 to 1, but 15 to 1 can be achieved using two sets of gears. For example, a 5 to 1 reduction ratio followed by a 3 to 1 reduction gives an overall step-down ratio of 15 to 1 (5 multiplied by 3 equals 15).

This is achieved in the drive mechanism shown in Figures 1.15 and 1.16. Figure 1.15 shows the initial stages of the drive mechanism, with an 8-tooth gearwheel on the motor driving a 40-tooth type on the second shaft. This in turn drives a third shaft via another 40-tooth gearwheel, giving a 1 to 1 drive ratio. Finally, an 8-tooth gearwheel drives a fourth shaft having a 24-tooth gearwheel.

This arrangement could be simplified somewhat by having the final 8-tooth gearwheel somewhere on the second shaft, and the 24-tooth gear wheel on the third shaft. The fourth shaft would then be unnecessary. However, in practice you will often find that one part of a mechanism tends to obstruct another part so that you can not do things exactly as you would wish. You may have to put in some additional parts in order to get things into place and performing in the desired fashion. Obviously this over-engineering should be avoided as far as possible.

Gear spacing

When dealing with gears you will not always find that the spacing from one gearwheel to the next is something nice and convenient. Lego have made things as easy as possible by keeping the radius of each gearwheel to exact multiples of 0.5 units. From the smallest to the largest the radii are 0.5, 1.0, 1.5, and 2.5 units. On the face of it there is no problem when using any of these, except when the 16-tooth wheel with its 1.0 unit radius is mixed with any of the others. The total spacing between the gearwheels is equal to the sum of the two radii, which is obviously not going to be an integer value if the 12-tooth gear is used with any of the others.

In reality matters are not as simple as that, because the normal Lego units do not apply when one beam is added on top of another. Two 8-tooth gearwheels have a total spacing of one unit, but due to the larger vertical units will fail to touch at all when used in two beams placed one on top of the other. Be prepared to add in some extra plates to get the correct spacing between gears, or to do some redesigning. Sometimes offsetting one gearwheel one unit sideways or vertically from the other one helps to get the spacing just right.

The larger gearwheels have holes that will take pegs. This is useful if the rotating wheel must operate a mechanism of some kind on each rotation. A protruding peg fitted to the wheel can be used to activate the mechanism.

Worm drive

The Robotics Invention System includes two worm gears, but you may have to search carefully through the packs of smaller parts to locate them. They

Fig.1.17 The two worm gears

are the two black spiral-shaped pieces (Figure 1.17). The two worm gears can be butted together to provide a single double-length gear, but you have to be careful to get them the match up correctly so that "you can not see the join" (Figure 1.18). If the spiral does not carry on correctly from one gear to the next, remove one gear, rotate it through 90 degrees, and refit it onto the shaft. If it still does not look right, repeat the process once or twice until the two sections do match up correctly.

A worm drive provides an output that is perpendicular to the drive shaft, as in the example of Figure 1.19. The most important point to note with this type of gearing is that the worm drive can drive the ordinary gearwheel, but not vice

Fig.1.18 The two worm gears can be combined into one long gear

versa. This is something of a limitation, but you can make "a virtue of a necessity". A worm drive effectively has a built-in brake, and it ensures that the load only goes where the motor directs it. If something is lifted via a worm drive there is no risk of it descending again as soon

as the motor is switched off! Of course, there may well be situations where it is essential for the drive to free-wheel in this way, and some other form of gearing must then be used.

Fig.1.21 *Bevel gears in action*

Fig.1.22 *A right angle drive using a crown gearwheel*

Fig.1.23 Two Lego racks

can sometimes be persuaded to work as conventional gearwheels with the standard gears. The crown gear is designed to be dual purpose, and it will happily work with the other gearwheels in a conventional arrangement, or to provide right angle gearing (Figure 1.22). The crown gearwheel can be on the drive shaft or the driven shaft. This method of right angled gearing works equally well either way round. You will notice that the 24-tooth gearwheel fitted on the motor in Figure 2.22 is not the usual grey variety, but is an alternative chunkier white plastic version that is included in the kit.

On the rack

A rack (Figure 1.23) is basically just a flat version of a gearwheel, and its purpose is to convert rotation into linear movement, or vice versa. The rack must be attached to something in order to perform a useful task, and plates, beams, etc., can be fitted on the underside. Two racks can be joined end to end by fitting them on top of an 8 x 1 beam or plate. There are four racks in the kit, and all four can be joined end to end if a really long rack (16 units) is required.

Fig.1.24 A simple rack and pinion mechanism

The rack must run in a groove to ensure that it does not skew or even slip away altogether from the driving gear wheel. The base and sides of the groove must be smooth, which gives a slight problem. Beams can be used as the sides of the groove, and they will provide reasonably low levels of friction. The base of the groove is a different matter. Beams and plates are knobbly and will not give the rack a smooth ride. Apparently the Lego system does include some smooth pieces ("tiles") that would probably "fit the bill" quite nicely, but there is a slight snag in that none of these are supplied in the Robotics Invention System. However, you can improvise with the parts that are supplied, and the girder style pieces are quite useful in this role (Figure 1.24).

Any of the normal gearwheels are usable with a rack, as are the crown gears. The 8-tooth gearwheel is the normal choice for this application though, since it gives more precise control. There are ten teeth on each rack, so it takes little more than one turn of an 8-tooth gearwheel to move the rack through its full travel. This is still better than the other gears though, with only one quarter of a turn being needed with a 40-tooth gearwheel for example. In most practical applications the rack mechanism must be preceded by reduction gearing even if the 8-tooth gearwheel is utilized. Alternatively, the rack seems to work quite well with a worm gear. With ten teeth per rack, it takes ten turns of a worm gear to move a rack by an amount equal to its length.

Differential

If you look through the various components in the Robotics Invention System you can not fail to notice an intriguing piece that comprises a 24-tooth gearwheel joined to a 16-tooth type. This is intended to form the basis of a differential, and it is used in conjunction with three bevel gears and two axles (Figure 1.25). One of the bevel gears fits on the spigot within the differential, and this must be fitted first. Then fit one of the other gears in place and fit its axle. Finally, fit the other gearwheel in place and thread its axle through the differential and into the gearwheel. This is a bit fiddly, but you should soon have everything assembled correctly.

Fig.1.25 The completed differential

The main use for a differential is when driving the rear wheels of a vehicle from a single motor and steering with the front wheels. In other words, the arrangement used in most full-size motor vehicles. If the rear wheels are fitted on a common axle there is a slight problem when cornering. The inside wheel takes a slightly shorter course than the outside wheel, and should turn more slowly. With the two wheels turning at the same rate there has to be some skidding of the tyres in order to get the vehicle around corners. This is probably not too important with small robots, but the differential gives you the opportunity to do things correctly. The drive can be applied to either the 16-tooth or 24-tooth gearwheel on the differential, and both axles will be driven. However, the differential gives the axles a degree of autonomy, and enables them to operate at different speeds when necessary. I will not try to explain how it works. If you make up a differential and experiment with it for a while you will soon see the secret of its success.

Pulleys and bands

Pulleys offer an alternative to gears where step-up or step-down drives are needed, or where you simply need to transfer power from point A to point B. A big advantage of using pulleys is that, unlike gears, you are not restricted to having a certain distance between the two shafts. When using the full-size "real thing" there can be any distance within reason between the two shafts. The drive belt is cut to suit the size of the pulleys and the distance between them. With the Lego system there are drive bands of various sizes, but the elasticity of the bands means that each one can accommodate a reasonable range of pulley sizes and spans.

Although pulleys and drive bands are in many ways more versatile than gears, it is only fair to point out that they do have one major disadvantage. This is simply that systems of this type are prone to slippage, and the elasticity of the drive bands can sometimes result in a lack of action initially as the system "takes up the slack" in the driving band. Particularly when it is loaded heavily, a pulley system may operate a bit erratically. This is another example of it being possible to "make a virtue of a necessity", with this slippage and elasticity providing some protection for the motor if the loading is excessive. With luck, severe overloading will result in the drive band slipping instead of the motor burning out.

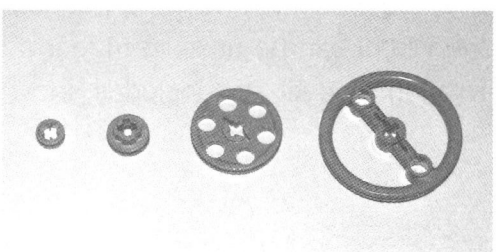

Fig.1.26 *The four sizes of Lego pulleys*

Fig.1.27 *A pulley drive having a step-down ratio of approximately 3.5 to 1*

There are four sizes of pulley included in the Robotics Invention System (Figure 1.26), and the drive ratio depends on the relative diameters of the two pulleys in the drive system. Relative to the smallest pulley, the other three have diameters of 1.5, 3.5, and 5.5. Using the smallest pulley and the second largest type, with the small pulley on the drive shaft, would therefore give a reduction ratio of 3.5 to 1 (Figure 1.27). Note that with a pulley system both shafts turn in the same direction, whereas the driven shaft goes in the opposite direction to the drive shaft when using gears.

Cams

The larger pulleys and the larger gearwheels, have holes to take pegs so that the wheel can be used to trigger a mechanism at a certain point on each rotation. The kit also includes a few cams, but only in one size. However,

Fig.1.28 A simple cam operated mechanism

there are four positions for the drive shaft, which help to give some variation in the amount and style of the movement. A cam is an irregular shaped wheel that is used to convert rotary motion into reciprocating movement. In other words, a rotating shaft produces up and down, backwards and forwards, or side to side movement. The simple example shown in Figure 1.28 uses a cam to produce an up and down action in a lever.

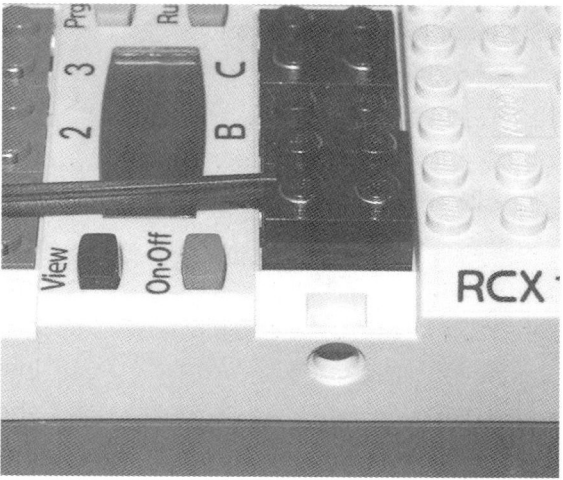

Fig.1.29 *If this method of connection has the motor operating correctly...*

The rest

It is not really practical to give a detailed description of all the parts in the Robotics Invention System, but most of the basic building blocks have been covered here. Wheels, tracks, flexible tubing, and many other parts are covered in the practical examples featured in chapters three to six. It is perhaps worth mentioning here that the motors should be connected to the RCX unit exactly as shown in the photographs in the subsequent

Fig.1.30 *...Then this method will have it turning in the wrong direction*

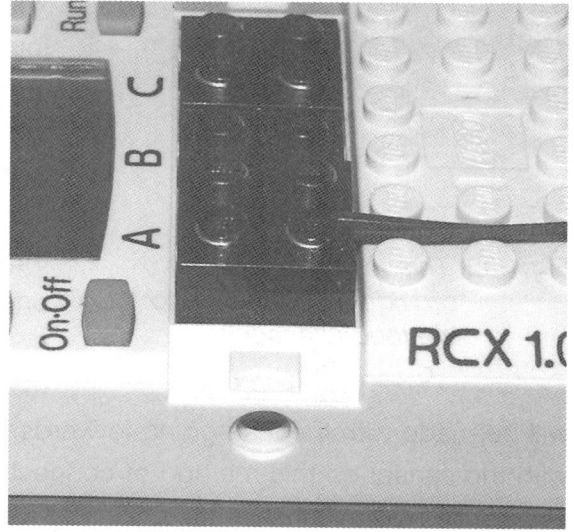

Fig.1.31 This method of connection is all right

Fig.1.32 This method does not provide a connection to the touch sensor

chapters. Whichever of the four possible orientations you use to connect the lead to the RCX unit or the motor, the motor will actually work. The potential problem is that it will turn the wrong way, sending your robot forwards when it should go backwards, and vice versa. For instance, if the method of connection shown In Figure 1.29 works properly, then the method shown in Figure 1.30 will always have the motor turning the wrong way.

The touch sensors are more accommodating, but you have to bear in mind that there are no terminals on the rear pair of nodules. You can make the connection to the four nodes at the middle and front of the sensor using any orientation, but the connection should only be made to the rear and middle nodules using the method shown in Figure 1.31. If the connection has the lead going to one side or the other, as in Figure 1.32, no connection will actually be made to the sensor. A connection will be made with the lead going forwards, but the lead could then interfere with the sensor so this method of connection should be avoided. None of this applies to the light sensor which has its lead permanently attached. The sensors can be connected to the RCX unit with any orientation incidentally.

Checking connections

It is worth bearing in mind that the RCX unit can display the readings from sensors, and this is useful for checking that the sensors are working correctly. With the RCX unit switched on, press the View button and the display will show readings from a sensor connected to input 1. Press the View button again to read input 2, and once more to read input 3. An arrowhead on the display indicates the input that is being monitored. A small triangle appears on the display against the appropriate input if it is being used to supply power to an active sensor such as a light type. In Figure 1.33 the display is showing the reading obtained from a light sensor on input 2. The arrowhead and the triangle are both showing against input 2, combining to give a filled-in arrowhead. Note that the range of numbers displayed

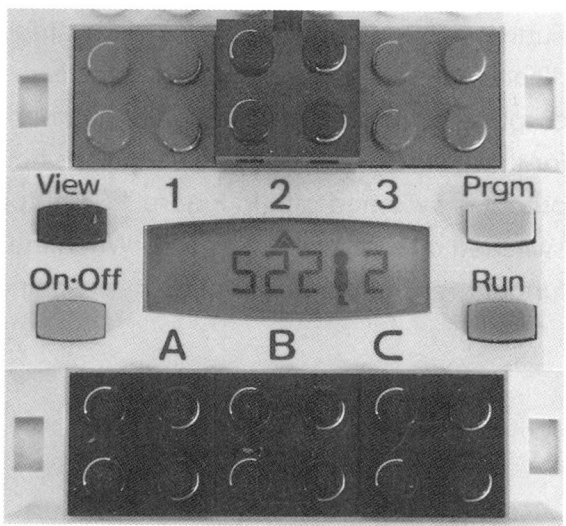

Fig.1.33 The display of the RCX unit can be used to monitor a sensor

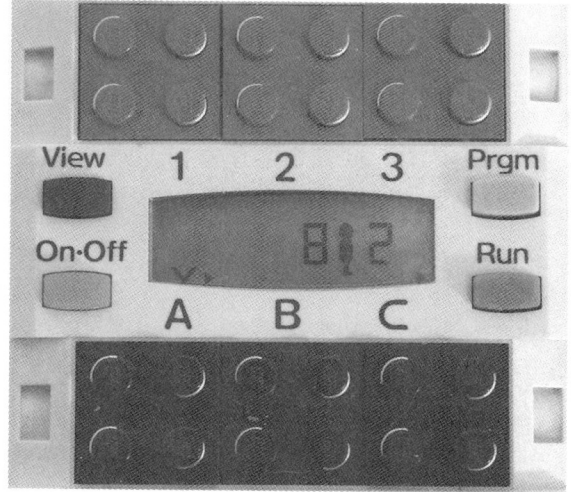

Fig.1.34 The RCX unit's display can also be used to monitor an output

when monitoring an input depends on the current operating mode of the input. This subject is covered in chapter 2.

Further presses show the status of the outputs, and a little dot appears on the display against any output that is active. The arrowhead indicates which output is being monitored, and the number in the display is the power setting for that output. The power settings are from 1 to 8, and 0 indicates that the output is switched off. In Figure 1.34 output A is being monitored, and it is switched on at maximum power. When the arrowhead is against output C, a further press of the View button cycles things back to the beginning again.

Using Spirit.OCX

RCX code

Amalgamating Lego with a microcontroller to produce "intelligent" robots was a great idea, but is one that had a potential stumbling block. In order to make the finished robots do something they had to be programmed. The Lego solution to this problem is RCX code, which enables non-programmers to easily program finished robots to do simple tasks. In fact it is possible to make your robots perform quite complex tasks using RCX code, but without the aid of a 50-inch monitor large programs in RCX code can become rather unwieldy! Also, RCX code does have one or two limitations, with the main criticism generally being its lack of variables. There is in fact a counter that can be used as a variable of sorts, and the instructions are arranged in such a way that variables can be largely avoided, but this is something of a limitation when you wish to do something clever.

For those who feel restricted by RCX code, or would simply prefer to use more conventional programming techniques, there are alternative programming languages available. A tour of Lego MindStorms related web sites will soon reveal some of these. It is actually possible to program the RCX unit using virtually any Windows programming language and the ActiveX control supplied with the Robotics Invention System. This file is automatically loaded onto your computer when you install the RCX software. The subject of ActiveX controls is a complex one, but the purpose of this particular control is to provide extra facilities to a programming language so that it can communicate with and program the RCX unit. Obviously a normal programming language does not have facilities to communicate with an RCX unit via the infrared link, and neither does it have commands such as switch off motor A, or read touch sensor 1.

The ActiveX control adds a set of instructions that give the same sort of control as RCX code, and it also provides a set of routines that enable the programming language to communicate with the RCX unit using the infrared link. For reasons that seem to be far from clear, the ActiveX control supplied with the Robotics Invention System is called Spirit, or Spirit.OCX to give it the full filename. The .OCX extension is the one used in all ActiveX filenames incidentally. Programs that need this file will probably be able to locate it themselves, but its default location is in the C:\Program Files\LEGO MIND STORMS\System directory.

Operating modes

When using Spirit.OCX and a programming language there are two modes of operation, which are the direct and indirect modes, which are also known as the immediate and delayed modes. The indirect mode is the one normally used with RCX code, where a program is downloaded to the RCX unit and then run within the RCX unit. In other words, you download a program to your robot which then runs the program, making it follow a line on the floor, do somersaults, or whatever.

With the direct method the program runs within the PC, which communicates with the RCX unit via the infrared link. When using the indirect mode the Windows programming language is little more than a means of downloading software to the RCX unit. You utilize the extra commands and facilities provided by Spirit.OCX, but most of the language's capabilities are left unused. A fortunate consequence of this is that you do not have to be a particularly expert programmer in order to program the RCX unit in this way. A good knowledge of the extra instructions provided by Spirit.OCX is clearly essential, but you can get by with a fairly rudimentary understanding of Visual BASIC, Delphi, or whatever language you are using.

Using the direct mode is very different, and a better understanding of the programming language is required. At the most basic level, the direct mode simply provides a means of manually controlling a robot. Using on-screen slider controls and buttons you control the direction of the robot, its speed and direction, etc. At a more advanced level, the infrared link is used to pass readings from the sensors to the computer, which then sends appropriate instructions back to the robot. This is similar to having a program running in the RCX unit, but the PC instead of the microcontroller in the RCX unit does the "thinking". In theory this enables more powerful control software to be used, because the super powerful Pentium processor in the PC rather than the simpler processor in the RCX unit controls the robot. In practice this extra computing power may not be of any great advantage, and it also has to be borne in mind that communication by way of the infrared link introduces small delays that could be problematic.

Another way of handling things is to have the robot send sensor readings back to the PC where they are displayed onscreen. The robot's operator can then respond to this information, and can even control the robot "blind" if the information from the sensors is sufficiently detailed.

VB6

In the space available here it is not possible to give details of using Spirit.OCX with a wide range of programming languages. We will concentrate on Visual BASIC 6, which is the obvious choice as it is by far the most popular programming language. The examples given here should also work with Visual BASIC 5, but note that this version has somewhat different menu structures to version 6. If you try these examples using Visual BASIC 5 it might be necessary to search through the menus to find some of menu items mentioned here. There are five versions of Visual BASIC 6, and any of them should be suitable for writing control programs for Lego robots. The three commercial flavours are the Standard version, the more expensive Professional Edition, and the even more costly Enterprise Edition. There is also an Academic version, but this is more or less the Professional Edition with different licensing conditions.

There is also a "free" version that is very occasionally given away on the cover discs of computer magazines, and it is also supplied with some beginners' books about programming Visual BASIC. This is virtually the full program, and it has full save facilities, etc., and can run any programs you write with it or load into it. Its main limitations are the lack of any online help and an inability to compile a program group that can be installed on a PC. Programs can only be run from within Visual BASIC 6, rather like running programs using a BASIC interpreter such as GW BASIC. This is not necessarily a major drawback in the current context, where you will probably not wish to compile and install lots of small programs on your PC anyway. It is probably best to run them from within Visual BASIC 6, which avoids having to compile and install numerous programs, and also makes it easy to implement changes to the programs. You only need to compile programs if they are to be distributed to people who do not have Visual BASIC 6.

VBA

It is perhaps worth mentioning VBA here, which is Visual BASIC for Applications. This is a program that is supplied with a number of Microsoft applications programs, including Word and some of the other Microsoft Office programs. It is also supplied with some other Windows programs, such as AutoCAD 2000. Although I have sometimes seen it suggested that there is no difference between Visual BASIC and VBA, this is far from the truth. The two programs have different purposes, and Visual BASIC is, of course, primarily intended for producing standalone programs. VBA on the other hand, is designed for adding new features to existing applications. This does not mean that VBA can not be used with Spirit.OCX to download programs to the RCX unit, etc., and it is indeed possible to use VBA in this fashion. It does mean that VBA can not be used in exactly the same way as Visual BASIC. VBA would not be my first choice for use with the Robotics Invention System, but if you have this on your computer but do not have Visual BASIC, it is certainly worthwhile giving it a try.

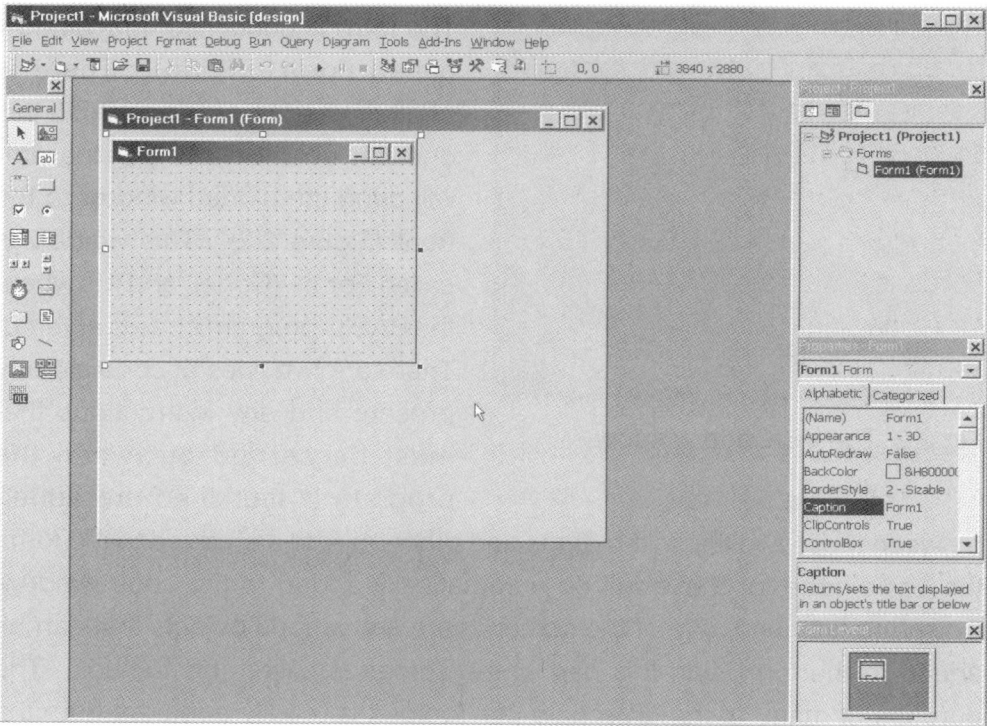

Fig.2.1 The screen should look something like this when VB is first run

Starting with VB

When Visual BASIC is first run you are presented with a window that offers various program types, but for our purposes the default Standard.EXE option will suffice. Simply press the Open button to select this. Once into Visual BASIC you will see several windows open on the screen, which should look something like Figure 2.1. The window covered with a grid of dots and called Form1 is the one that you use to design the window in which the compiled program will run. The compiled program will run in a window the same size as the form. If a control button is placed on the form an identical button will appear at the same position when the program is run. In this case the programs may never be compiled and installed as a standalone program, and may simply be run from within Visual BASIC. However, when the program is run a new window containing the control buttons, etc., will appear.

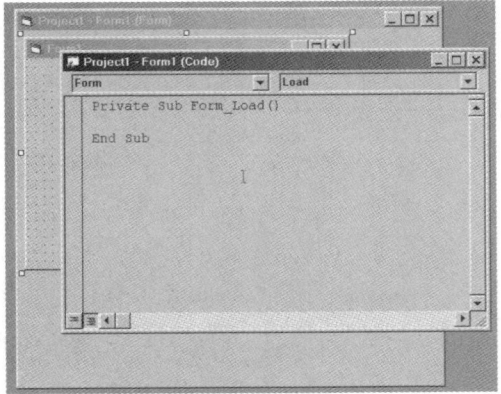

Fig.2.2 The VB Code window

The form is contained in a window called Form1 (Form). There is another window hidden behind this one, and double clicking on the form will bring the hidden window to the front (Figure 2.2). This window is called Form1 (Code), and it is where the BASIC program code is entered. There are two lines of code already present and any instructions that must be carried out when the program is launched are added between these two lines. If buttons and other objects are added to the form, the basic code for these will be automatically added to the code window. Down the left-hand side of the screen there are various objects that can be added to the form, and this part of the screen is called the Toolbox. The purpose of many of the objects in the Toolbox is readily apparent from the icons used. However, placing the cursor over any of the icons will bring up a brief description of its function.

The component we will need most is the Command Button object. If it is not already fully visible left-click anywhere on the form to bring it to the fore. Then left-click on the Command Button icon and place the cursor over the form. Left-click on the form and then drag a rectangle of suitable size for the control button. Once the button has been created it can be dragged to any required point on the screen, and it has eight handles that permit it to be easily resized. The button will be labelled Command1 by default, but it is easily relabelled. In the right-hand section of the screen there is a Properties window, and this shows the current properties of the selected object. Left-clicking on the form or an object on the form will make it the current item and bring up its details in the Properties window, but the new button will already be the current object. If you look in the Properties window you will find a property called Caption, and beside this it will say Command1. In order to edit this name left-click on

Command1 and then change it to whatever you require using normal text editing methods. In this case the caption is changed to DOWNLOAD, which should immediately result in the label on the button changing in sympathy.

Spirit.OCX?

As explained previously, in order to write code that can be sent to the RCX unit the Spirit.OCX control is required. This object is loaded onto the form in exactly the same way as command buttons and other objects. There is a slight problem though, and if you look at the objects in the window down the left-hand side of the screen it is unlikely that Spirit.OCX will actually be there. It will probably not be loaded by default, and you must add it manually. To do this, activate the Project menu and then select Components. This brings up a list of additional objects, and in the alphabetical list you should find one having a name that starts with the word Lego. It will probably be called something like PBrick Control rather

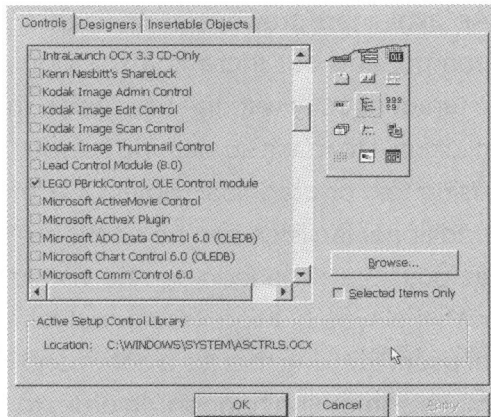

Fig.2.3 The list of installable components

than Spirit, but this is the component you require. Left-click on the box beside its entry to place a tick there (Figure 2.3), and then operate the Apply and OK buttons. The very distinctive Lego logo should then appear in the palette of objects.

If for some reason Spirit.OCX is not in the list of objects you must select the Browse option and then find it. As pointed out previously, by default it will be found in the C:\Program Files\LEGO MINDSTORMS\System directory. Having found Spirit.OCX in the file browser, left-click on its entry to select it and then operate the Open button to add it to the list. It can then be added into the palette of objects in the usual way.

It should not be necessary to go through this procedure each time Visual BASIC is run, and having loaded Spirit.OCX once it should be automatically loaded each time Visual BASIC is launched. Unfortunately, things can sometimes go slightly awry, and this mainly seems to happen when the PC has one or more versions of VBA running in addition to Visual BASIC. The new control usually loads into VBA correctly, but unfortunately it has to be added to Visual BASIC each time it is run. If there is an easy solution to this problem I am unaware of it. You simply have to keep loading the control, or save a form with this control added and use it as your starting point.

As explained previously, Spirit.OCX is added onto the form like any other component, but there is an obvious difference in that it does not actually interact directly with the user when the program is running. You can not left-click on it to make something happen, and it does not display any information using text or graphics. Spirit.OCX must still be added to the form so that its additional facilities can be used, but it is advisable to make its icon as small as possible. The icon can be visible or hidden when a program is run, and with this type of thing it is normal for it to be hidden. Otherwise there is a risk of users thinking that it's a command button, and that left-clicking on it should make something happen. In order to hide the icon, first select Spirit.OCX and then look at its properties in the Properties window. Select the Visible property, left-click on the button that appears to its right, and then select False. Note that the icon will still be visible on the form, but it will not appear on the screen when the program is run.

Adding code

So far we have a Visual BASIC form with Spirit.OCX added, together with a control button (Figure 2.4), but the button does not actually do anything. Double clicking on the control button will bring up the code window, and the cursor will be placed between the two lines of code for the button that have been added for you by Visual BASIC. This is where the code for the control button must be placed. Notice that the first line of code ends with Click(). This means that the code you enter will be performed when the control button is left-clicked. There are other options available, but you will normally require the code for a button to be performed only when the button is left-clicked. This is certainly all we require here, so there is no need to alter the default setting, but this is easily done via the Properties window for the button.

Before going on to consider the main commands available using Spirit.OCX we will consider a simple program that demonstrates the

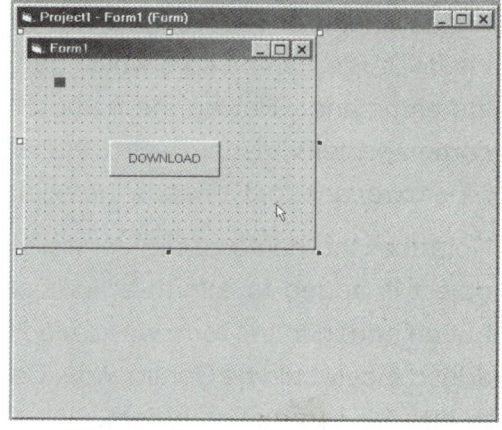

Fig.2.4 The Spirit.OCX icon must be present on the form, but only needs to be small

basic principle of downloading software. We are getting slightly ahead of ourselves here, but this program will enable you to check that everything is working correctly, in addition to illustrating some fundamental points. It simply switches on the motors connected to output A and C, waits two seconds, and then switches them off again. Add the extra lines of code so that the complete routine for the control button looks like this:

```
Private Sub Command1_Click()
Spirit1.InitComm
Spirit1.SelectPrgm 1
Spirit1.BeginOfTask 0
Spirit1.On "02"
Spirit1.Wait 2, 200
Spirit1.Off "02"
Spirit1.EndOfTask
End Sub
```

Even the least observant will have noticed that each of the new lines start with Spirit1., and this is because the code in all these lines relies on Spirit.OCX. There are no lines in pure Visual BASIC. The name of the ActiveX control is added ahead of each command, with a full stop between the two so that Visual BASIC can be sure which part is the name of the control and which is the command. Having the name of the ActiveX control added ahead of a command tells Visual BASIC that the command is not part of its normal repertoire, and that it needs the additional facilities of the ActiveX control.

Note that the name of the control is Spirit1, and not just Spirit. When an object is added to a form a number is always added after its name. The button added to the form was called Command1, and if further buttons were added they would be Command2, Command3, etc. Incidentally, it is possible to edit the name of an object such as this by left-clicking on it to select the object, and then editing the name in the Properties window. With only two or three objects on a form it is probably not worthwhile doing so however. You should certainly not confuse matters by changing the name of Spirit1.

The first of the new lines simply initialises the infrared link to the RCX unit. This does not actually result in anything being sent to the RCX unit, and its purpose is to get the PC ready to use the infrared link. The next line specifies a program number, which is actually a block of memory in the RCX unit that is reserved for program storage. There are five of these memory blocks and five program numbers available. On the RCX unit's display these are numbered from 1 to 5, but in programs you must use numbers from 0 to 4. This anomaly crops up quite frequently when writing programs for the RCX unit, and is not uncommon in general computing. It stems from the fact that we like to number things from one upwards, but the way in which computers handle numbers results in a starting point of zero. There are ways around this problem, but it is not that difficult to remember that an adjustment of one is needed in some values. In this case we are using program number 1, which is program 2 as far as the RCX unit's display is concerned.

Tasks

Line three of the additional code starts a task, and in this case it is the only task. There can be up to ten tasks numbered from 0 to 9, and they can run simultaneously if necessary. For example, there could be two touch sensors and a light type, with a separate task continuously monitoring each one. This type of thing is normally organised on the basis of a main task which does any initial setting up that is required and then starts the other tasks. Where this multitasking is not required, there must still be one task, because the RCX unit only understands programs that are organised as one or more tasks. Hence our simple motor control example is the main task, and does not call any additional tasks.

Having indicated the beginning of a task, the next lines actually make that task do something. In this case the motors on outputs 0 and 2 are switched on. The outputs of the RCX unit are marked A, B, and C, but when using Spirit.OCX they are referred to as outputs 0, 1, and 2 respectively. A Wait instruction then results in the program doing nothing for two seconds. The first figure in this instruction (2) indicates that the source of the delay is the second figure (200). The delay is set with a resolution of 10 milliseconds, or one hundredth of a second if you prefer. A value of 200 therefore provides a delay of two seconds. An Off instruction then switches off both motors. Finally, an EndOfTask instruction tells the RCX unit that the end of the program is reached. You must be careful to include this instruction at the end of each task, because programs will not operate properly if they are omitted.

In order to try this program, activate the Run menu and then select Start. The program should then run, and it should look like the form but without the grid of dots (Figure 2.5). Set up the infrared transmitter in exactly the same way as you do when transferring RCX code to the RCX unit, and then operate the DOWNLOAD button. This should result in the green light in the transmitter switching on for a few seconds while the program is transferred to the RCX unit. The program number that appears on the RCX unit's display should be 2, together with four zeros on the left-hand section of the display. If all is well, pressing the Run button on the RCX unit will result in the motors running forwards for two seconds. If not, carefully

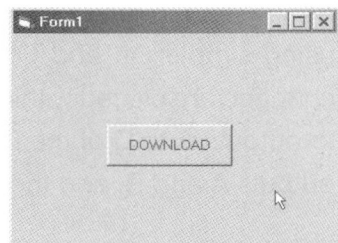

Fig.2.5 The simple test program in action

check through the program for errors and then try again. With programs it helps to bear in mind that if you get everything right it will work. If you do not get everything just right, then a program will certainly not work. To terminate the Visual BASIC program left-click on the cross in the top right-hand corner of the program window.

The commands

A wide range of instructions is available for programming the RCX unit, and it is not possible to give a detailed description of every single instruction here. This would be pointless anyway, since some of the available commands are of limited use for general programming, and are not the type of thing you will use initially. Accordingly, we will concentrate on the commands that are needed for everyday programming of your robots. There is no documentation for Spirit.OCX supplied with the Robotics Invention System, but there is a document in Adobe PDF (portable document format) called the Software Developer Kit that is available from the Lego MindStorms web site. This provides a description of every command plus some general information, and it is something that anyone who will be using Spirit.OCX should download. If possible it should also be printed out so that it is handy for reference purposes whenever it is required.

Motor On/Off

As we have already seen, one or more motors can be switched on or off using simple On and Off commands. The numbers of the outputs to be changed are placed within double quotation marks. Output 0 to 2 correspond to outputs A to C of the RCX unit. The first of these examples will switch on outputs A and B, and the second will switch off outputs B and C:

```
Spirit1.On "01"
Spirit1.Off "12"
```

SetFwd/SetRwd

The direction of motors is not set using the On command, but is instead set using separate forward and reverse instructions. These are respectively

SetFwd and SetRwd. They are used much like the on and off commands, with the outputs to be affected being specified in double quotation marks. These two examples set the motors on outputs A and C to go forwards, and the motor on output B to go in reverse:

```
Spirit1.SetFwd "02"
Spirit1.SetRwd "1"
```

Float

In addition to On and Off commands there is also a Float instruction. This switches off the affected motors, but it is not quite the same as the ordinary off command. With the float command the motors are stopped in a free running mode. If you run the simple test program provided previously, when the end of the two seconds is reached the motors stop almost instantly. The motors are not truly locked in position, but turning the shafts is quite difficult. The motors are stopped in what is termed brake mode. If the Off command is changed to an equivalent Float type, the motors will continue to spin for a short time after they have switched off, and the spindles will then be much easier to turn. This example sets outputs A and B to the float mode:

```
Spirit1.Float "01"
```

AlterDir

This command changes the direction of the specified motors. Unlike SetFwd and SetRwd this command is not setting a specific direction. Instead, it is simply making the new direction the opposite of whatever it happened to be before the command was issued. In most situations it is better to use the commands that set a specific direction, as this avoids the possibility of things getting out of kilter. This example reverses the direction of the motors on outputs A and C:

```
AlterDir "02"
```

Wait

This is another instruction that was used in the simple test program provided earlier. It simply provides a pause in the program, and it is most commonly used to keep one or more motors switched on or off for a specified period. It can simply be used to provide a delay for the period specified directly in the command, as in our earlier example, but the delay can be specified in other ways. The way in which the two parameters in this command function is shared with some other commands, so it is worth looking at this aspect of things in some detail.

The first parameter sets the source of the delay value, and there are 17 potential sources for the commands that use this system. However, most of these sources are inapplicable to most commands, and others are not the type of thing you would need very often, if at all. Probably the two most commonly used sources are constants and variables, which are selected using values of 2 and 0 respectively. A constant, also known as a literal, is simply the number given in the next parameter of the command. In this example, the first parameter is 2, meaning that the delay value is the constant given in the second parameter:

```
Spirit1.Wait 2, 100
```

The literal value is 100, giving a delay of one second. The allowable range of values in a constant depends on the command in use, but for a Wait instruction it can be any integer (whole number) from 1 to 32767. This gives a maximum delay of almost five and a half minutes, which should be more than adequate for the vast majority of practical applications..

Variables are something we will cover in more detail as and when necessary, but a variable is just a value stored in a memory location. In the case of the RCX unit there are 32 memory locations specifically set aside for storing variables, and these are numbered from 0 to 31. This command has zero as the first parameter and 12 as the second, and the delay will therefore be controlled by the value stored in variable number 12.

```
Spirit1.Wait 0, 12
```

The delay this provides depends on the value stored in variable number 12, and the program must have placed a suitable value in this variable before the Wait instruction is performed. This version of the Wait instruction is only used where the delay time will need to be changed while the program is running, and this is easily accomplished by altering the value stored in the variable.

Random

The Wait instruction produces a random delay if the first parameter is 4. The second parameter then sets an upper limit for the delay. For instance, this example produces a random delay of between one hundredth of a second and 10 seconds:

```
Spirit1.Wait 4, 1000
```

This may not seem to be very useful, but there are practical applications for randomness. If you go exploring and find yourself going round in circles or otherwise going over the same old ground time and time again, you will realise what you are doing and break out of the pattern. A rover style robot running a simple control program has no way of knowing whether it is wandering far and wide, or is stuck in a fixed search pattern. Adding a random element can help to avoid getting your robots "stuck in a rut". A random element is also an essential feature of some simple games and novelties that can be implemented using the Robotics Invention System.

SetPower

You are not limited to simply switching the motors on and off, and it is possible to adjust the power fed to the motors. There are eight power levels from 0 (minimum) to 7 (maximum). The three parameters in this command start are first a list of the motors to be changed, then the source of the new power setting, and finally the power setting itself. This example would therefore set the motors on outputs A and C at power level 3:

```
Spirit1.SetPower "02", 2, 3
```

In this version the power of outputs A and C is set to the value stored in variable 9:

```
Spirit1.SetPower "02", 0, 9
```

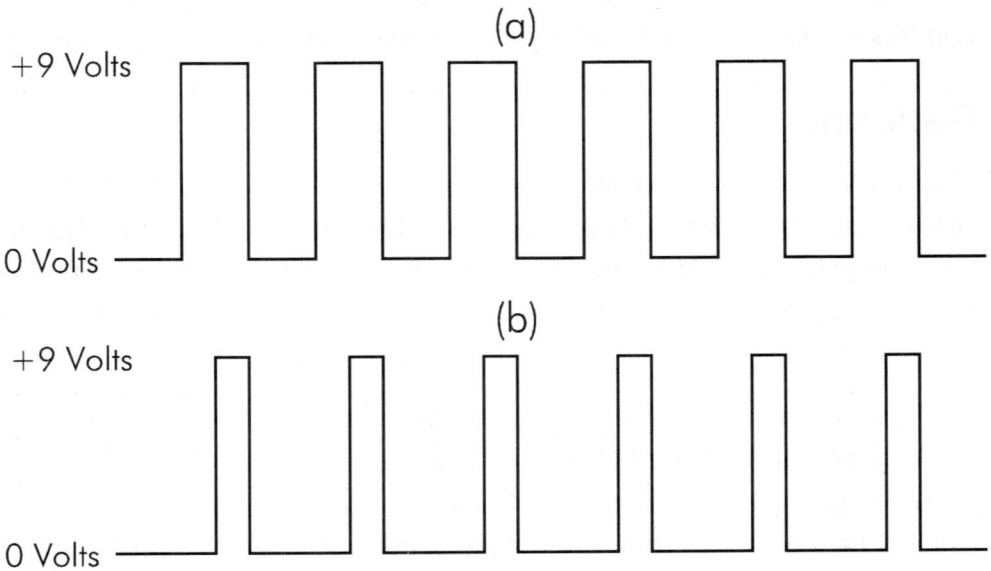

Fig.2.6 The output power is varied using a form of pulsed control

There are some important points to bear in mind when using power settings. Setting a power level does not in itself switch a motor on. This must be done previously or subsequently using the On command. Similarly, setting a power level of 0 does not switch a motor off. A power level of 0 gives a very low output power, but it does not cut off power to the motor altogether. Do not confuse power levels with speeds. Reducing a motor from full power to half power does not necessarily reduce its rotation speed by half. If the motor is only loaded very lightly there will be very little reduction in speed, but if it is loaded heavily it might actually grind to a halt. Often the only way to find the power settings that give the desired result is to use trial and error.

The obvious way of controlling the output power is to simply vary the output voltage. Unfortunately, in practice this method does not work too well when used to control DC electric motors. It varies the speed well enough, but at lower powers the motor has a tendency to stall, and once stopped it is reluctant to start again. The method of power control used by Lego is a form of pulsed control. This operates by sending pulses at full power to the motor so that it is working at full power for a percentage of the time, and is switched off for the rest of the time. If the pulses are sent at too low a frequency the motor will judder along, but provided they are sent at a high enough frequency the motor will run reasonably smoothly.

The motor responds to the average output power, and in the waveform of Figure 2.6(a) the output voltage is switched on for 50 percent of the time, so the motor will receive half power. In the waveform of Figure 2.6(b) the output is only switched on for 25 percent of the time, so the motor is operated at one quarter power. The point of this system is that the pulses at full power resist any natural tendency for the motor to stall. Also, if the motor should stop, or it is being started at a low power level, the pulses will usually nudge a reluctant motor into action. This method of control tends to produce very noticeable "buzzing" sounds from a motor when it is used at low power levels.

SetVar

We have seen that some commands can use the values stored in variables, but this is only of use if there are ways of setting and manipulating the values stored in variables. Setting a variable to a specific value is performed using the SetVar command, and this has three parameters. The first selects the variable to be altered, the second selects the source, and the third is the value itself or the exact source. In this example variable number 4 is set to a value of 1023:

```
Spirit1.SetVar 4, 2, 1023
```

The valid range of values for a variable is integers from –32768 to +32767. However, make sure that the values of variables are appropriate to any commands they will be used in. The Wait command can only be used with positive numbers for example, and output powers can only be from 0 to 7.

SumVar

This command adds the specified value to the contents of a variable. The result of the addition is stored in the variable. It operates in much the same way as SetVar, with the three parameters selecting the variable to be altered, the source of the value to be added, and the exact source. In the first example

100 is added to the contents of variable 20, and the result is stored in variable 20. If variable 20 originally contained 70, it would contain 170 after this instruction had been completed. The second example adds the contents of variable 8 to variable 22 and stores the result in variable 22. If variables 8 and 22 respectively contained values of 55 and 200, once the instruction had been completed they would contain 55 and 255.

```
Spirit1.SumVar 20, 2, 100

Spirit1.SumVar 22, 0, 8
```

In practice SumVar is unlikely to be used with large values, but if it is, make sure that the results of this command stay within the maximum permissible value of 32767. Adding (say) 25000 to 20000 will cause an overflow, and will not produce an answer of 45000.

SubVar

This is the subtraction equivalent of the SumVar command. It subtracts the indicated value from the specified variable, and stores the result in the variable. The first example subtracts 54 from variable 12. If variable 12 contained a value of 20 initially, it would contain −34 after this command had been performed. In the second example the contents of variable 23 are subtracted from the value stored in variable 16. The answer is stored in variable 16, and the value in variable 23 is unaffected. If variables 16 and 23 were respectively at values of 100 and 45 before the command was executed, they would be at values of 55 and 45 afterwards.

```
Spirit1.SubVar 12, 2, 54
Spirit1.SubVar 16, 0, 23
```

MulVar

DivVar

These two commands operate like SumVar and SubVar, but provide multiplication and division. The first example multiplies the value in variable 12 by the value in variable 22. If these two variables respectively contained 15 and 5 before the command was issued, they would contain 75 and 5 afterwards. The second example divides the value stored in variable 18 by 6. If variable 18 contains 39 before the command is executed, it will contain 6 afterwards. Where necessary, the answer is rounded down so that it is always an integer.

```
Spirit1.MulVar 12, 0, 22
Spirit1.DivVar 18, 2, 6
```

SgnVar

Using this command it is possible to test the sign of the value stored in a variable. It has three parameters, which are the variable used to store the result of the test, the source, which will usually be another variable, and the number of the variable to be tested. These are the three values that can be produced as a result of the test:

Value	Meaning
0	The variable is set at zero
1	The variable is set at a positive value
−1	The variable is set at a negative value

This example tests the value in variable 23 and places the result in variable number 4:

```
Spirit1.SgnVar 4, 0, 23
```

If the value in variable 23 was 18 when the command was issued, it would remain at this value afterwards, and variable 4 would be set to a value of 1.

AbsVar

This command places in the specified variable the absolute value of the number that is tested. In other words, it always produces a positive result regardless of whether the source value is positive or negative. For instance, this example would test the contents of variable number 24 and place the result in variable 20:

```
Spirit1.AbsVar 20, 0, 24
```

If variable 24 contained –101 before this command was issued, it would still contain –101 afterwards, and variable 20 would be set at a value of 101. If variable 24 contained 101, then variable 20 would still be set at 101.

AndVar

OrVar

These operate much like the other arithmetic commands, and they are used to provide bitwise And and Or logical operations on the two values involved. Bitwise operations enable some bits of a binary number to be read or altered without reading or affecting other bits. This is not usually of great importance when dealing with the RCX unit, where there is individual control over each output, and there is no difficulty in reading just one of the three inputs.

Flow control

Every program language requires commands that deal with flow control. On the face of it there is no need for any form of flow control, and a program can simply start at the beginning, go through a list of commands, and then stop. In practice it is hard to find applications where this sort of linear programming applies. When dealing with robots it is unusual for the robot to simply go through a series of commands and then stop. Even if the robot must perform some very simple production line task, having performed an action once, it must then go on to repeat it over and over again. This simply entails going back to the beginning again once the end of the program has been reached. This repeating of a program, or part of a program, is known as looping. In some applications the robot must perform one task or another, depending on the data from a sensor. It is known as a branch when a program goes one way or another depending on the result of a decision-making instruction.

Loop

EndLoop

In order to repeat one or more commands a certain number of times it is just a matter of using the Loop command. Actually a loop of this type requires two commands, which are Loop at the beginning of the loop and EndLoop at the end. The Loop command has two parameters that control the number of times the commands within the loop are performed. These point to the source of the value and the actual value in the usual way. The EndLoop command does not have any parameters. This routine will switch output A on and off 11 times:

```
Spirit1.Loop 2, 11
Spirit1.On "0"
Spirit1.Off "0"
Spirit1.EndLoop
```

This routine has the same effect, but the value stored in variable 27 controls the number of loops:

```
Spirit1.Loop 0, 27
Spirit1.On "0"
Spirit1.Off "0"
Spirit1.EndLoop
```

Sometimes you need the loop to carry on indefinitely, such as when monitoring a sensor for example. This is achieved by telling the loop to repeat zero times, as in this example, which will switch output A on and off indefinitely:

```
Spirit1.Loop 2, 0
Spirit1.On "0"
Spirit1.Off "0"
Spirit1.EndLoop
```

If

EndIf

On looking through the list of available commands when using Spirit.OCX you might come to the conclusion that there is no way of reading the sensors. There is a Poll command, but this is only usable as an immediate command that enables the program running in the PC to read a sensor. The normal way of handling a sensor is to read its value and place the result in a variable. A decision-making instruction then does one thing or another, depending on the value read from the sensor. When using Spirit.OCX things are normally streamlined somewhat, and the decision-making instruction can directly read the sensor. Hence there are no separate instructions for reading the sensors. This aspect of things is handled by the flow control commands. Note though, that it is possible to do things the roundabout way if preferred, and the SetVar can set a variable to the value read from a sensor. A value of 9 is used to select a sensor as the source, and values of 0 to 2 respectively select sensors 1 to 3. This command would therefore set variable 10 to the value read from sensor 2 (input 3):

```
Spirit1.SetVar 10, 9, 2
```

Probably the most common way of reading a sensor is to use an If...EndIf routine. One or more commands are placed between the If and EndIf instructions, and these commands are only performed if a certain condition is met. There are five parameters in the If command, but none in the EndIf type which simply marks the end of the routine. The first two parameters point to a value using the standard type of source and precise source method. The final two values point to a second value using the same method. This leaves the middle parameter to define the type of test being performed, and these are the available options:

Number	Test Performed
0	>
1	<
2	=
3	<> (not equal)

Suppose that we required output A to be switched on if input 2 returns a reading that is greater than 50, this routine would provide the desired result:

```
Spirit1.If 9, 1, 0, 2, 50
Spirit1.On "0"
Spirit1.EndIf
```

The functions of the five parameters in the If command break down thus:

9   Read Sensor

1   1 (input 2)

0   If the returned value is greater than...

2   the constant value...

50   50, then perform these program lines…

In this case there is only one line of code to perform, which turns on output A. An EndIf instruction then terminates the routine. Of course, the first value in the comparison does not have to be a reading from a sensor, and the second value can be something other than a constant. It is quite acceptable to have an If instruction compare the values in two variables for instance.

Else

The If instruction is actually part of a standard If…Then…Else program structure, but the Else element is optional. In the previous example output A is switched on if input 2 returns a value that is more than 50, but the output is not switched off if the returned value is 50 or less. Output 2 is simply left in whatever state it had previously if the returned value is not greater than 50. This version, which utilizes the If...Then…Else structure, will switch off the motor if the returned value is not greater than 50:

```
Spirit1.If 9, 1, 0, 2, 50
Spirit1.On "0"
Spirit1.Else
Spirit1.Off "0"
Spirit1.EndIf
```

The commands between the If and Else instructions are performed if the returned value is greater than 50, and those between the Else and EndIf instructions are performed for returned values of 50 or less. In this case there is only a single instruction in each section of the routine, but there can be numerous commands if necessary.

While

EndWhile

An If routine tests a sensor (or whatever) a single time and then moves on. It can be combined with a loop to perform the comparison a certain number of times, or it can be used with an infinite loop to keep on testing indefinitely. There is an alternative method of repeated testing in the form of While and EndWhile instructions. These form what in conventional computing is called a do...while loop. In other words, an instruction or list of instructions is performed while a certain condition is met. There are five parameters in a While instruction, and these operate in the same way as the five parameters in an If instruction. This routine will keep switching output A on and off while the reading from input 1 is less than 75:

```
BeginOfTask 0

Spirit1.While 9, 0, 1, 2, 75

Spirit1.On "0"

Spirit1.Off "0"

Spirit1.EndWhile

EndOfTask
```

As we saw with the simple test program described previously, programs downloaded to the RCX unit must be organised into one or more tasks. The BeginOfTask and EndOfTask instructions are used to indicate the beginning and end of each task. Note that these instructions do not actually result in anything being sent to the RCX unit, but they are needed by Spirit.OCX to enable it to correctly organise the data that is downloaded to the RCX unit. These instructions are not required when using the RCX unit in the immediate mode. Note that BeginOfTask is followed by the number of the task (0 to 9), but no number is required in an EndOfTask instruction.

GoSub

BeginOfSub

EndOfSub

A subroutine, or subprogram as it is also known, is a series of commands that are called up by the main program as and where necessary. A subroutine is normally used where the same task must be performed at various places in a program. Rather than repeating the routine at every point in the program where it is needed, it is defined as a subroutine, and then called from the main program at the appropriate places. Once the subroutine has been performed the program goes back and continues where it left off. The subroutine is called using the GoSub command, and this has one parameter, which is the number used to identify the subroutine. Up to eight subroutines can be used, with numbers from 0 to 7. With many programming languages it is possible for one subroutine to call another. Some memory (called the Stack) is used to keep track of the jumps from one subroutine to another, so that the program can always go back to the right place once a subroutine has been completed. The RCX unit does not have a Stack though, and calling one subroutine from another is not allowed. The subroutine starts with a BeginOfSub instruction and finishes with an EndOfSub type. The BeginOfSub instruction has one parameter, which is the identification number of the routine. The EndOfSub command has no parameters. In this example subroutine 5 is called, and simply switches output C on and off:

```
Spirit1.GoSub 5

...

...

Spirit1.BeginOfSub 5
Spirit1.On "2"
Spirit1.Off "2"
Spirit1.EndOfSub
```

On the face of it there is little difference between a subroutine and a task. The all-important difference is that tasks provide multitasking. In other words, if you call up three tasks one after the other, they will all be performed at once. If three subroutines are called up in the same way, one routine will be completed before the next one is commenced. Tasks are used where you need things to happen simultaneously, such as having two or three sensors continuously monitored. Subroutines are used where you simply wish to use the same routine at various points in a program.

StartTask

StopTask

As their names suggest, these two commands can be used to stop and start a task. These commands carry only one parameter, which is the number of the task to be halted or started.

Timers

The RCX unit has four timers that have 100 millisecond (0.1 second) resolution. There is only one command that controls the timers, and this is the ClearTimer command. This simply resets the specified timer to zero, and after being reset it immediately resumes counting. The timers are numbered from 0 to 3. The timers can be read by certain commands if a value of 1 is used as the type of source parameter. For example, this command would set variable 10 at the value read from timer 2:

```
Spirit1.SetVar 10, 1, 2
```

The While and If commands can also be used to read the timers. This example will keep outputs A and C switched on for ten seconds:

```
Spirit1.ClearTimer 0
Spirit1.While 1, 0, 1, 2, 100
Spirit1.On "02"
Spirit1.EndWhile
Spirit1.Off "02"
```

First Timer 0 is cleared, and then a While loop keeps turning on outputs A and C until the value in Timer 0 reaches 100. Once outputs A and C are switched on, further On instructions obviously have no effect, but they do no harm either. These are the functions of the five parameters in the While loop:

1   While Timer…

0   0

1   is less than…

2   the constant value…

100   100, do this…

The timers have a resolution of 0.1 seconds, so waiting for the count to reach 100 gives the required 10 second delay before the final line switches off the motors.

PlayTone

When you start using the RCX unit you soon become aware that it has a simple sound generator facility that can generate tones. You can utilize this facility via the PlayTone instruction, which has two parameters. The first is the required frequency in hertz, and the second is the duration of the tone in hundredths of a second. The frequency value must be in the range 1 to 20000, which gives coverage of more than the full audio range. The time value must be from 1 to 255. This command would therefore produce a note a 440Hz for two seconds:

```
Spirit1.PlayTone 440, 200
```

If a program includes a list of notes, the list is stored in the RCX unit so that one note is completed before the next one is commenced. It is therefore possible to use this instruction to play simple tunes. Page 110 of the Software Developer Kit includes a chart that shows the frequencies required for a wide range of musical notes.

PlaySystemSound

In addition to simple tones, six predefined sounds can be produced using the PlaySystemSound command. This instruction has a single parameter, which is the number of the sound to be played (0 to 5). These are the available sounds:

Number	Sound
0	Key click
1	Beep…beep
2	Sweep upward in frequency
3	Sweep downward in frequency
4	Buzzing sound
5	Rapid upward sweep in frequency

This command will provide the buzzing sound, which lasts a second or two:

```
Spirit1.PlaySystemSound 4
```

SetWatch

The RCX unit includes a 24-hour clock facility, or a "software watch" in Lego terminology. The watch can be set to the correct time using the SetWatch command, which is followed by two parameters. These set the hours and the minutes. This command would therefore set the watch to 15:22:

```
Spirit1.SetWatch 15, 22
```

The watch can be read using SetVar, If, and While instructions, and its source number is 14.

InitComm

CloseComm

As we saw with the example program given at the beginning of this chapter, the InitComm command has to be issued before communication with the RCX unit is possible. This command does not send any data to the RCX unit, but instead sets up the appropriate serial port to operate properly with the infrared transmitter. Once communications has been completed, the CloseComm command can be used to restore things to normal operation. It is not essential to use the CloseComm command if the serial port is only used with the infrared transmitter, but it must be used before switching back to use the port with another device.

On the face of it there is no problem in simply adding the CloseComm command at the end of the program used to download software to the RCX unit. In practice this does not work because the CloseComm instruction comes into effect before the software has been fully downloaded. A serial port is relatively slow, so a PC uses a small amount of memory, called a buffer, to store serial port data until it can be transmitted. Presumably the program data is stored in the buffer and then the CloseComm command closes down the infrared transmitter before all the data has been sent. A delay could be added ahead of the CloseComm command, but there is no way of knowing how long each program will take to transmit. However, this method should work provided you are fairly generous with the amount of time allowed. A simple alternative is to have an extra control button, which issues a CloseComm command and then closes the program used to download the software. The light in the infrared transmitter switches off once the software has been downloaded, so you just wait for this light to go out and then operate the extra button to close communications and terminate the program.

SendPBMessage

This command is used to send data from the RCX unit via the infrared link. This normally means sending data to the PC, but it is possible to have two RCX units communicate via the infrared link. This command can also be used to simply produce a brief burst of infrared "light" from the transmitter, which is something that is exploited by one of the robots in chapter 3. This instruction is followed by two parameters, which are the type of source, and then the exact source. This command will transmit a value of 127

```
Spirit1.SendPBMessage 2, 127
```

Values for transmission must be integers in the range 0 to 255. Incidentally, the PB part of this command's name simply stands for Programmable Brick. This is a term that is often encountered when dealing with Lego MindStorms kits, and it is just another name for the RCX unit.

Sources

As we have already seen, several commands use parameters that are given via an indirect means, with the first parameter giving the type of source (constant, variable, etc.), and the second pointing to the source more precisely (the actual value of a constant, the number of a variable, etc.). This is a full list of the available sources, but note that not all of these sources are usable with some commands. The charts in the Software Developer Kit show the sources that are applicable to each command.

Number	Source	Second parameter
0	Variable	Number of variable (0 - 31)
1	Timer	Number of timer (0 - 3)
2	Constant	Value to be used (–32768 to +32767)
3	Motor status	Motor number (0, 1, or 2)
4	Random	Maximum value for number (1 - 32767)
5	Tacho counter	Counter number (0 or 1)
6	Tacho speed	Speed number (0 or 1)
7	Motor current	2
8	Program number	0
9	Sensor reading	Number of sensor (0, 1, or 2)
10	Sensor type	Number of sensor (0, 1, or 2)
11	Sensor mode	Number of sensor (0, 1, or 2)
12	Sensor raw	Number of sensor (0, 1, or 2)
13	Sensor Boolean	Number of sensor (0, 1, or 2)
14	Watch	0
15	PB Message	0
16	AGC	0

Some of these are the type of thing that you will need to use time and time again, such as constants and variables, while others have to be regarded as non-essential. Note that sources 5, 6, 7, and 16 only apply to the Lego CyberMaster kits, and not those that utilize an RCX unit.

The commands described here are the main ones that are needed for general programming. There are some others detailed in the Software developer Kit, but they are mostly only applicable to the Lego CyberMaster kits.

VB shortcut

Putting Spirit1 ahead of every command that utilises Spirit.OCX can be a bit tedious even when entering short programs into Visual BASIC. With longer programs it can become very irksome indeed. Fortunately, Visual BASIC provides a shortcut in the form of its With facility. Simply add With Spirit1 ahead of a block of instructions that use Spirit.OCX, and End With after the last of these instructions. The Spirit1 prefix can then be omitted from each instruction. The test program provided earlier could thus be reduced to this:

```
Private Sub Command1_Click()

With Spirit1

.InitComm

.SelectPrgm 1

.BeginOfTask 0

.On "02"

.Wait 2, 200

.Off "02"

.EndOfTask

End With

End Sub
```

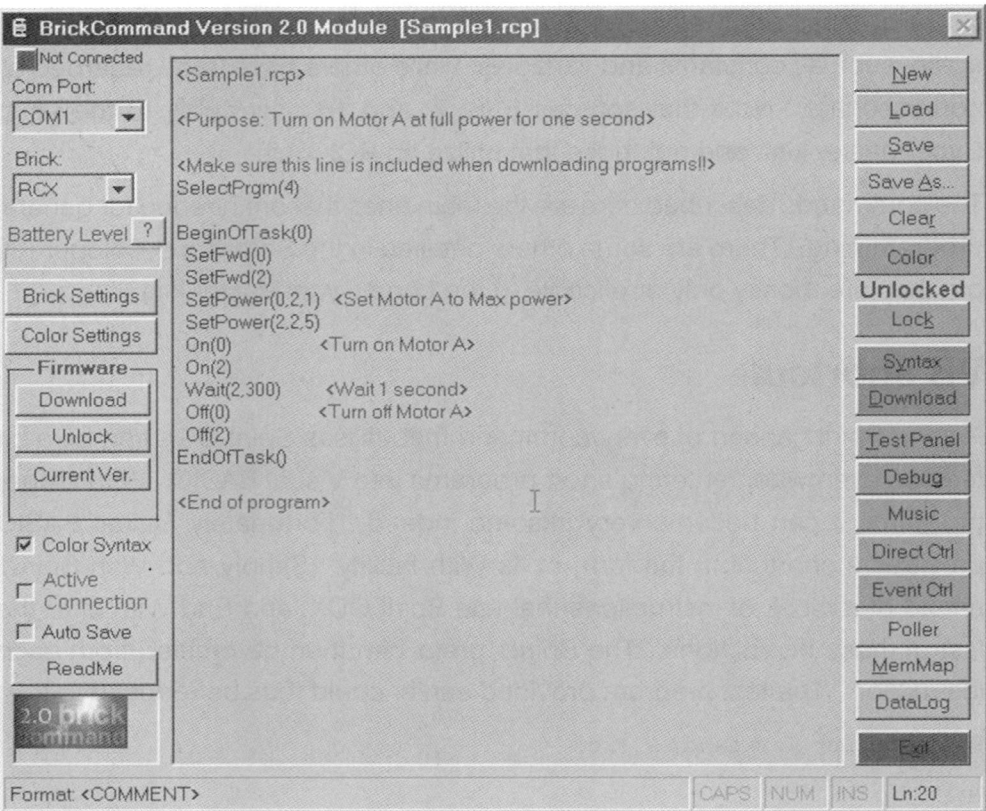

Fig.2.7 The BrickCommand program can be used to download programs

VB alternatives

If you do not have access to Visual BASIC there are alternative means of downloading your programs to the RCX unit, and a tour of Lego MindStorms related web sites will soon find a few of these alternatives. There are complete programming languages such a NQC (not quite C), and software that enables Visual BASIC style programs to be downloaded to the RCX unit. In order to download the programs featured in this book it is clearly one of these Visual BASIC alternatives that is required. Probably the best known of these, and the only one I have tried, is BrickCommand (Figure 2.7) which is at version 2.0 at the time of writing this. This program is freeware, so there is no charge for using it.

One or two points have to be borne in mind when using BrickCommand with programs written for visual BASIC. The first of these is that it is only suitable for programs that are downloaded to the RCX and then run on the RCX. It is not suitable for Visual BASIC programs that control the robot in immediate mode. This is due to the fact that BrickCommand

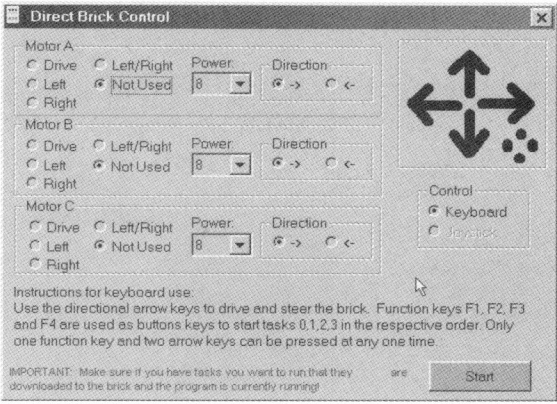

Fig.2.8 The BrickCommand remote control facility

does not have the range of Visual BASIC instructions, controls, graphics, and other features that are used with this type of program. It is designed for downloading software to the RCX unit, which it does very well. BrickCommand does have an immediate mode that provides a sort of remote control facility that is usable with many robots (Figure 2.8).

When using BrickCommand there is no need to use InitComm and CloseComm, as the program handles all this for you. Neither is it necessary to put Spirit1. at the beginning of every command, because the program knows that all the commands require Spirit.OCX. Parameters are placed within parentheses, with no space between the command word and the first parenthesis. The parentheses are still needed even if an instruction does not have any parameters. The simple test program provided earlier in this chapter would therefore be reduced to this if used with BrickCommand:

```
SelectPrgm(1)
BeginOfTask(0)
On(02)
Wait(2, 200)
Off(02)
EndOfTask()
```

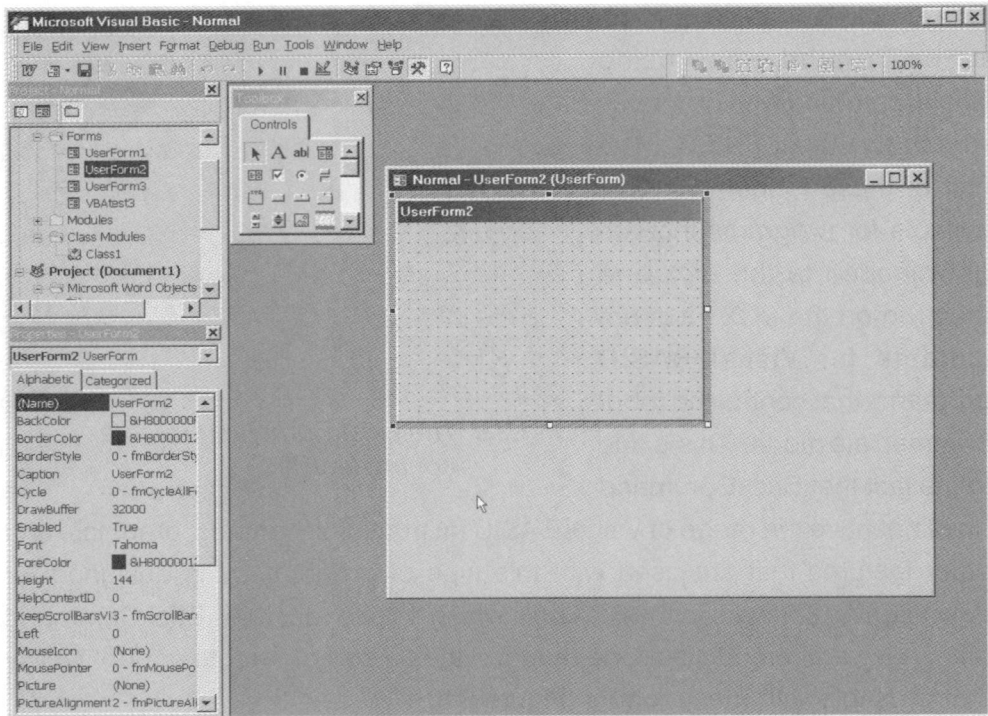

Fig.2.9 VBA looks similar to VB and offers much the same facilities

Using VBA

It may take a little searching to find VBA in a suitable host application, but in AutoCAD 2000 and Microsoft Word it can be found by activating the Tools menu, and then selecting Macros and Visual BASIC Editor. This should bring up a screen something like Figure 2.9. If there is no form present on the screen by default, activate the Insert menu and select User Form. If the Lego logo is not present in the Toolbox, activate the Tools menu and select Additional Controls. Find Spirit.OCX in the list provided, and left-click on the box beside its entry in order to add it to the Toolbox. It appears as Spirit Control on my PC, but it might be listed under a different name. Operate the OK button to go back to the VBA editor, and the icon for Spirit.OCX should then appear in the Toolbox.

Things are then much as before, with a button being added to the form, along with the Spirit.OCX control. Double-click on the button to bring up the code window so that the program for the button can be added. When the program has been entered activate the Run menu and select Run Sub/User Form to test it. The program window will then appear, and activating the button will download the program to the RCX unit. A Properties window for the form will appear if you click on the form to make it active. You can then change its name and caption to something more suitable, and save the form, either by operating the appropriate button or via the File menu. The new form will then be added to the list in the window towards the top left-hand corner of the screen. A form is closed by left-clicking on the cross in the top right-hand corner of its window, and retrieved by double-clicking on its entry in the list of forms.

Using VBA is admittedly rather more limited than programming with the "real thing", but it can certainly be used to try out some programs, and at zero cost if you already have an application that includes VBA.

Lightbots

Light work

The Lego Robotics Invention System is supplied with two touch sensors and one light type. The latter is probably the more useful of the two types and opens up some interesting possibilities. It can be used to produce robots that will follow lines on the floor, seek out light or dark, and with the aid of the RCX unit's infrared transmitter it can detect objects before a robot hits them rather than afterwards. The rover type robot described here (Figures 3.1 and 3.2) is simple but tough, and is an excellent test-bed for experimenting with the light sensor. It has been designed to travel relatively slowly so that it is easy to see exactly what is happening when it is used with various programs, but the gearing is easily changed if desired. In its initial form the robot has the light sensor positioned to permit operation as a line follower.

The basic robot consists of three main sections. First a chassis complete with wheels is constructed, and then the RCX unit and a motor assembly are fitted to the chassis. The light sensor is then fitted in the appropriate position, depending on the application of the robot. The rear wheels are driven by separate motors to permit steering in the standard Lego robot fashion, and the front wheels have separate axles in order to give reasonable manoeuvrability. The lack of front wheel streering does somewhat limit the manoeuvrability of the robot, but it can still perform a useful range of tricks. A tracked robot having improved manoeuvring skills is described later in this chapter incidentally.

Fig.3.1 and Fig.3.2 The completed Lightbot robot

Step 1 (Figures 3.3 - 3.5)

The chassis is based on two symmetrical side pieces that are each constructed from a 10 x 1 beam, a 12 x 1 beam, and a 16 x 1 beam. Each side section is held together using four black "rivets", and you must be careful to get these in the right holes or they might block the axles later.

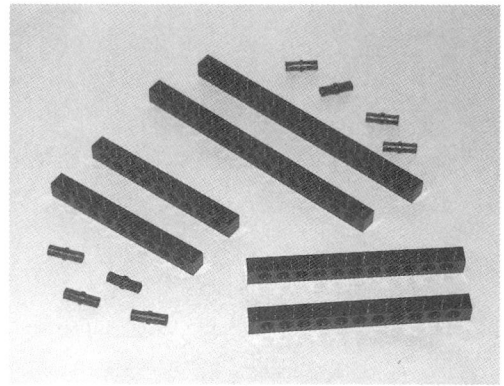

Fig.3.3 The components for the side pieces

Step 2 (Figures 3.6 and 3.7)

Next, add a number of grey plates to what then becomes its upper surface to join the two sides of the chassis. These plates also provide a more secure platform for mounting the RCX unit. A 6 x 2 plate is added across the rear, with an 8 x 1 plate just ahead of this. Fitting two 2 x 1

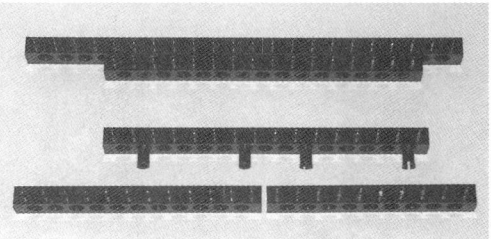

Fig.3.4 Assembling the side pieces

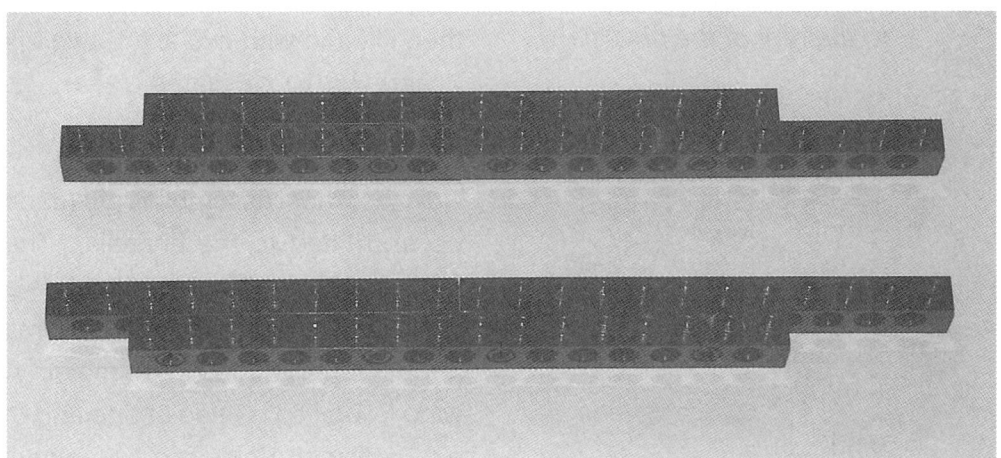

Fig.3.5 The completed side pieces for the chassis

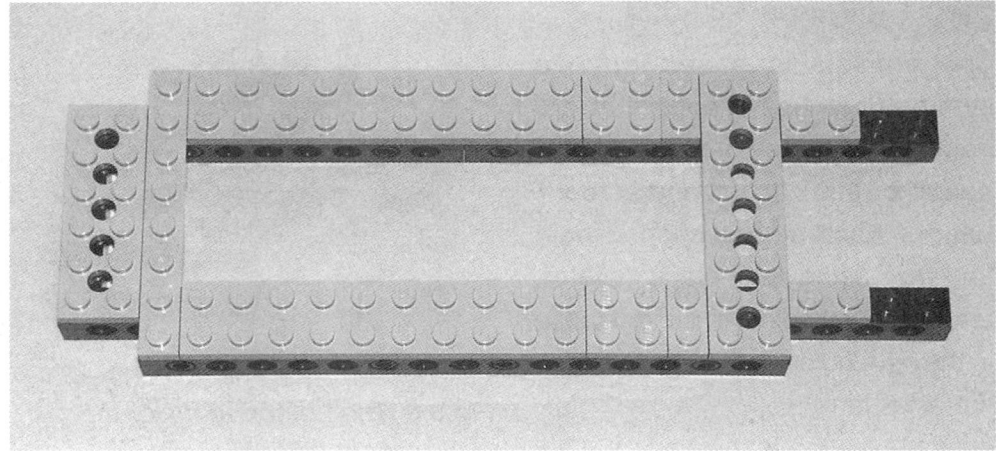

Fig.3.6 The plates added to the top of the chassis

*Fig.3.7 The two blocks added
to the rear of the chassis*

black blocks under the 6 x 2 plate, as far to the rear as possible, will help to strengthen the rear of the chassis. An 8 x 2 plate is added across the front section of the chassis, with two 2 x 1 plates immediately ahead of this. The gaps between the front and rear plates are then filled in with two 2 x 1, two 2 x 2, and two 10 x 2 plates.

Step 3 (Figures 3.8 - 3.10)

Next the wheel assemblies are made up and fitted to the chassis. The front wheels are the type that has a yellow hub and heavily treaded tyres about 45 millimetres in diameter. A wide fixing clip is fitted at one end of a 47-millimetre axle rod, the wheel is added, and then a second clip is

*Fig.3.8 The parts for the front
wheel assemblies*

Fig.3.9 The chassis with the front wheels added

added. A third clip is used to fix the assembly in place on the chassis. Of course, two of these wheel assemblies are required, one each side. Note that these wheels are not symmetrical, and they are presumably meant to be used with the concave sides facing outwards. The front wheels are mounted as far forward as possible on the section

Fig.3.10 Another view of the front wheels

of the chassis that is two beams wide. This leaves part of the chassis protruding beyond the front wheels, and where necessary this can take sensors.

Fig.3.11 The chassis with the wheels fitted

*Fig.3.12 The parts for the rear
wheel assemblies*

Step 4 (Figures 3.11 - 3.14)

The rear wheel assemblies are similar, but use the wide wheels about 50 millimetres in diameter and axle rods about 63 millimetres long. As before, two wide clips are used to fix the wheel at one end of the axle, but a 40-tooth gearwheel is then added alongside the second fixing clip.   These wheels have an indentation on one side and a protrusion on the other. Results look neater with the indentations facing outwards and the outer fixing clips partially inside the wheel hubs, but it seems to be possible to use them either way round. The completed rear wheel assemblies are added to the chassis on the section that is two beams wide, but not as far to the rear as possible. Fit the wheels

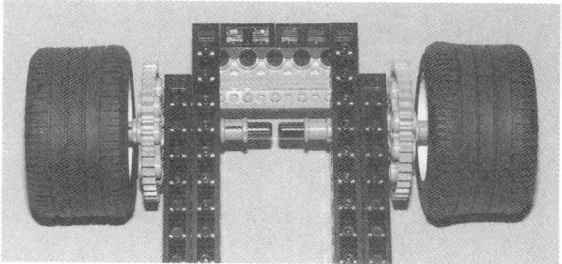

Fig.3.13 The rear wheels fitted to the chassis

one set of holes forward from this position, using a wide clip to hold each assembly in position.

Fig.3.14 Another view of the chassis with all the wheels fitted

Fig.3.15 The completed chassis and wheels assembly

Step 5 (Figure 3.15)

Fit two 6 x 2 plates, two 8 x 2 plates, and one 10 x 6 plate on the underside of the chassis to strengthen it.

Step 6 (Figure 3.16)

Fit the RCX unit on top of the chassis. The front of the RCX unit should be flush with the two 2 x 1 grey plates at the front of the chassis.

Step 7 (Figures 3.17 - 3.20)

The motor assembly is produced by first fitting two yellow 4 x 2 plates across the tops of the motors to fix them together. Small (8-tooth) gears are fitted to each motor, or 12-tooth gears can be used if slightly higher speed with less torque is preferred. The motor assembly can be fitted on the chassis in this form, but may have a tendency to come adrift. It can be mounted more securely by making up two mounting blocks that fit into the slots at the rear of

Fig.3.16 The RCX unit in position on the chassis

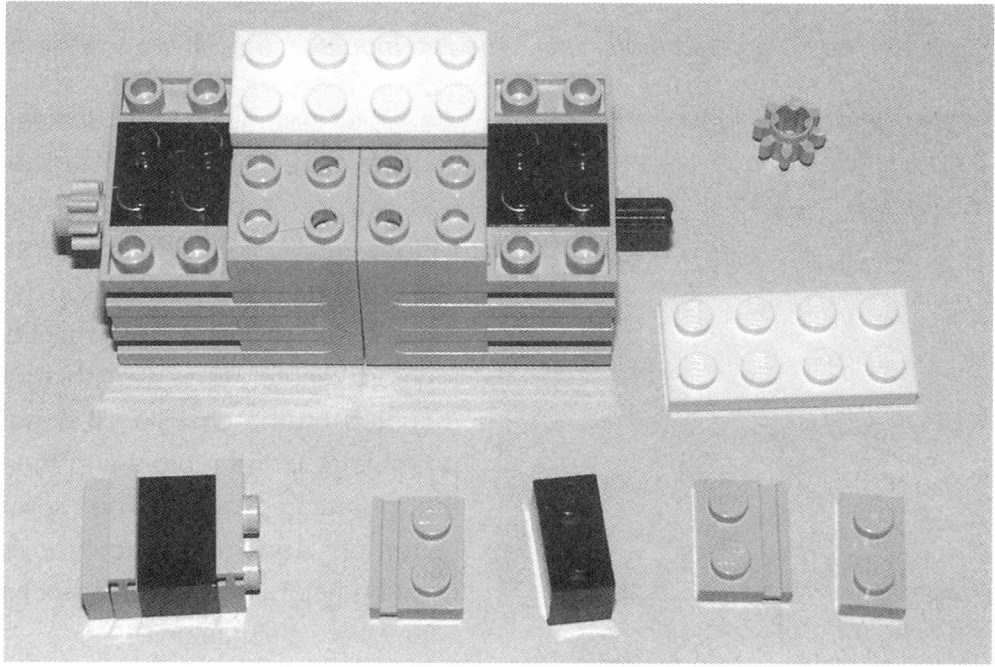

Fig.3.17 The part completed motor unit showing all the parts required

Fig.3.18 The completed motor unit. The two mounting blocks are a loose fit

Fig.3.19 12-tooth gears are also usable on the motors

the motor assembly. Each of these consists of a 2 x 1 block, a 2 x 1 plate, and two 2 x 1 plates having the flanges that fit into the motors. Note that the flanges are a loose fit in the motors, and this assembly only fits together securely once everything is mounted on the chassis. It is not possible to fit the motors and then the mounting blocks. The motor assembly and mounting blocks must be held together and fitted as if they were a single unit.

Fig.3.20 The motor unit fitted onto the rear of the chassis

Step 8 (Figures 3.21 - 23)

To complete the unit the two motors are connected to outputs A and C of the RCX unit using two short leads. Make sure that the connector blocks have the same orientation as those shown in the photographs, or the motors may not turn in the appropriate directions. The light sensor is mounted on the underside of the RCX unit at the front using a grey right angle plate (2 by 2 per section), and it is connected to input 2 of the RCX unit.

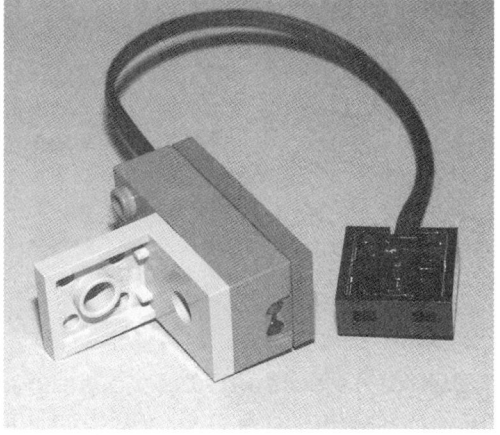

Fig.3.21 A right-angle plate is used to mount the sensor

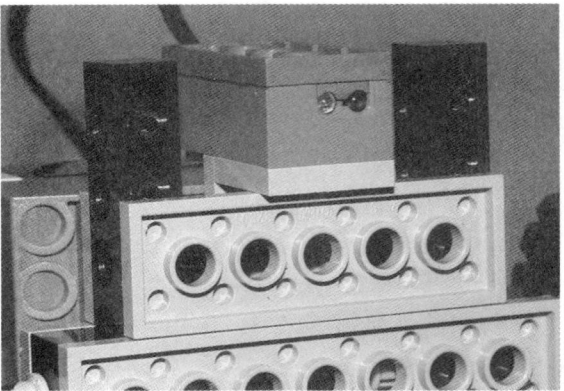

Fig.3.22 The light sensor is aimed down at the ground, and has only a few millimetres clearance

Fig.3.23 A view from above of the completed robot. The motors connect to outputs A and C, and the light sensor connects to input 2

Inside lane

This simple initial program enables Lightbot to follow an oval or circular course by following a black line on the ground. The obvious test track to use is the one provided with the Robotics Invention System, but it is obviously not too difficult to make your own using a large sheet of paper or cardboard. The better the contrast between the backing material and the line, the easier it is to get Lightbot to follow the line.

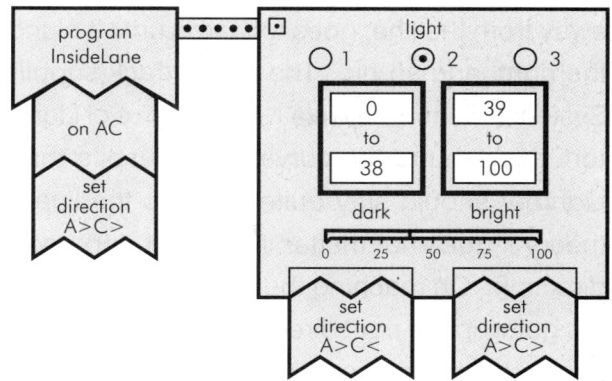

Fig.3.24 *The RCX code to make the robot follow the inside edge of a black line*

Getting the robot to follow the line on the inside of the oval is extremely easy, and requires only a very short program in RCX code (Figure 3.24). The main program only has two "bricks", which switch on the motors connected to ports A and C, and set the direction of the motors to forwards. Attached to the main program there is a light sensor "brick" set to port 2. This has just one programming block to tell Lightbot what to do when the sensor detects darkness, and another single block to provide control when a high light level is detected. When lightness is detected Lightbot simply has to move forward, so this "brick" sets A and C to the forward direction. When darkness is detected Lightbot must turn, and the direction in which it turns determines whether the line is followed in a clockwise or counter-clockwise direction. In Figure 3.24 motor A (on the left-hand side) is kept running forward, but motor C is reversed. This turns Lightbot to the right when the sensor goes over the black line, giving the clockwise tracking inside the line.

Overshoot and slippage

Following the line using this method is very much a matter of trial and error. Lightbot goes forwards until the line is reached, it then turns to the right to get away from the line, goes forwards until it blunders into the line again, turns to the right, and so on. The paper track supplied with the Robotics Invention System is quite small relative to the size of Lightbot making progress somewhat tortuous around the curves, but despite this very crude method of control Lightbot should stay quite close to the line once it has been detected. In theory it does not matter if Lightbot approaches the line going in the wrong direction. On reaching the line it should keep nudging round to the right until it is going in the right direction. In practice this may not always work, because some overshoot occurs when the black line is detected. There is a small gap between the line being detected and Lightbot actually turning to the right. This is partially due to a less than instant response time of the electronics, but is probably due in the main to mechanical constraints. An electric motor can not instantly switch from full speed in one direction to maximum speed in the opposite direction. If the line is crossed during the changeover, Lightbot is free and will "make a run for it". A slightly thicker line would totally avoid this problem. Using a faster robot would make matters worse, and for this application a slow but precise robot gives the best results.

Results might be good with the program set to produce clockwise or counter-clockwise movement around the track, but Lightbot will probably work much better in one direction than the other. This seems to be due to slippage on the paper track, that is something less than easy for the tyres to grip. The problem occurs where one tyre grips the track much better than the other does. Results are actually very good provided the tyre having the greater grip is the one that goes into reverse to turn Lightbot. If the other tyre has much greater grip there is a tendency for Lightbot to keep going straight on when it reaches the line, with the wheel going in reverse just skidding on the track. There are programming and mechanical methods of counteracting this problem, and these are discussed later in this chapter. There should be no problem of this type when robots are used on a surface that enables the tyres to get plenty of grip.

Another sort of slippage can cause problems, and this is where the paper track is place on a smooth surface such as vinyl or laminate flooring. You can sometimes find that as a robot moves in one direction the paper track moves in the other. This does not necessarily prevent a robot from performing properly, but it certainly does not seem to help. If the paper track starts sliding around it should be held in place while the robot performs, or it can be temporarily glued in place using something like Bostik Blu-Tack.

Viewing readings

By default, any sensor value from 51 to 100 is deemed to be light, and any value from 0 to 50 is dark. In practice this transition point might not give good results, and when the prototype was used with these values it tended to just keep on going when it reached the line. A transition point at readings of around 35 to 40 gave much better results. The best setting probably varies substantially from one sensor to another, so how do you determine the best light and dark ranges for your Lightbot?

When using sensors other than simple on/off types such as a touch sensor it pays to bear in mind that the RCX unit can be used to display the readings returned from a sensor. Switch on the RCX unit and then press the View button twice. This should place an arrow marker just below the figure 2 above the display, indicating that the displayed readings are from sensor 2. By placing the sensor over the line and white areas of the paper track you can see some typical light and dark readings. The ideal transition point is about half way between these two readings. For example, suppose that dark readings are around 34 while readings from light areas of the track are about 52. The average of 34 and 52 is 43 (34 + 52 = 86, 86/2 = 43), and good results should therefore be obtained with a dark range of 0 to 43, and a light range of 44 to 100. Once a robot is basically "up and running" you can always try some "fine tuning" of the threshold point to see if results can be improved. However, the light threshold level does not seem to be critical, and any value between the light and dark values should give good results.

VB program

With a task as simple as following a line there is probably no point in resorting to anything other than RCX code, but simple tasks of this type provide an

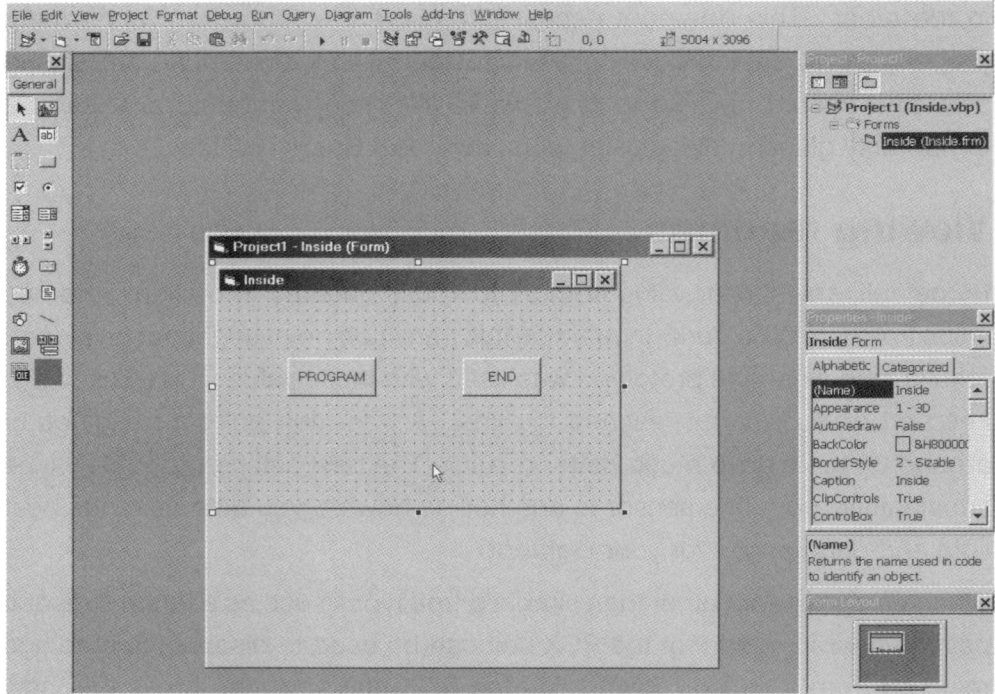

Fig.3.25 The Spirit control and two buttons are needed on the form

excellent introduction to using Spririt.OCX and a programming language such as Visual BASIC. We will therefore consider Spirit.OCX versions of most of the simple RCX programs provided in this book. If you wish to try the Visual BASIC equivalent of the line-following program provided previously, start Visual BASIC in its normal EXE mode and place onto the form the Spirit.OCX control and two buttons. Caption one button "PROGRAM" and the other "END". This should give a form similar to that shown in Figure 3.25. Double-click on the "END" button and then add two lines of code so that its section of the program is like this:

```
Private Sub Command2_Click()
Spirit1.CloseComm
End
End Sub
```

All this does is to close the communications channel to the serial port and end the program once the code for the RCX unit has been downloaded.

Return to the form and double-click on the "PROGRAM" button. Add to its basic entry so that its section of the program looks like this:

```
Sub Command1_Click()
Spirit1.InitComm
Spirit1.SelectPrgm 1
Spirit1.BeginOfTask 0
Spirit1.SetSensorMode 1, 4, 0
Spirit1.SetSensorType 1, 3
Spirit1.On "02"
Spirit1.SetFwd "02"
Spirit1.StartTask 1
Spirit1.StartTask 2
Spirit1.EndOfTask

Spirit1.BeginOfTask 1
Spirit1.Loop 2, 0
Spirit1.If 9, 1, 0, 2, 40
Spirit1.SetFwd "02"
Spirit1.EndIf
Spirit1.EndLoop
Spirit1.EndOfTask

Spirit1.BeginOfTask 2
Spirit1.Loop 2, 0
Spirit1.If 9, 1, 1, 2, 41
Spirit1.SetRwd "2"
Spirit1.EndIf
Spirit1.EndLoop
Spirit1.EndOfTask
End Sub
```

Although the Spirit.OCX version of the program looks very different to the RCX code program, it is actually very similar. The program looks large when compared to the RCX version, but you have to bear in mind that in the Spirit.OCX version you need more than just the raw commands. In the RCX program you selected a light sensor by simply dragging the appropriate icon into position. A few more clicks of the mouse selected the right port for the sensor and adjusted threshold settings. Using Spirit.OCX it takes several program lines to set the sensor type, mode, and port, and to set the threshold levels. In RCX code the program structure is largely handled for you, and there is no need to put in blocks to handle loops and branches. Programs that utilize Spirit.OCX do require lines to get the program structure and flow correct. Hence this program has far more lines than there are blocks in the version written in RCX code. However, this version is not really any more complex than the original.

The program consists of three sections. The first of these (Task 0) does some initial setting up, calls the other two tasks, and then ends. The other two tasks are infinite loops, one acting when darkness is detected and the other becoming active when high light levels are sensed. The main program starts by opening communications to the infrared link, and selecting a program number. In this case program number 1 is selected, which is actually program 2 in the RCX's numbering system, but you can obviously use any valid program number here. Next the sensor mode is set, and the first number selects port 1, which is input 2 in RCX terminology. The next value is 4, which sets the input to percentage mode. In other words, readings will be from 0 to 100. The final value is a dummy one that is of no relevance with this program. Having set the sensor mode, the type of sensor is then selected. The first value selects the correct input port and then a value of 3 is used to indicate that the sensor is a light type. Then the two motors are switched on and their direction is set to forwards. Finally, Task 1 and Task 2 are called, and the main task is terminated.

Task 1 detects when a certain light threshold is exceeded, and then sets both motors to the forward direction so that Lightbot goes forward in a straight line. The Loop instruction has a loop value of 0 to make this routine loop indefinitely. The If instruction has these five parameter settings:

9   read sensor…

1   on port 1 (2)

0   if the reading is greater than…

2   the constant value…

40   40, then do this…

When the reading is more than 40 only one instruction is performed, and this sets both motors to go forwards. The If structure is then terminated and the routine is looped back to the beginning again.

Task 2 is much the same, but the If instruction and the command following it have been changed. The modified If instruction has these five settings:

9   read sensor…

1   on port 1 (2)

1   if the reading is less than…

2   the constant value…

41   41, then do this…

The program line that follows this sets motor 2 into reverse and turns Lightbot to the right. Therefore, for light values of 41 or more Lightbot goes forward, and for readings of 40 or less it turns to the right, giving the desired action. With any program that uses multitasking, whether it is in RCX code or any other language, it is essential to make sure that the various tasks work in harmony. If the various tasks each handle only their particular section of the robot there is no risk of conflicts occurring, but often there will be two tasks controlling the same section of the robot. That is certainly the case here where one task sets motor 2 in a forwards direction and the other sets it into reverse. Provided the threshold values in the If instruction contain suitable values, only one or the other will try to set the motor's direction at any one time. Get things wrong and operation of the robot will be unpredictable and probably erratic. If your multitasking robot throws a fit it is worthwhile checking to see if the multitasking has been implemented correctly!

Testing

There is probably no point in compiling this sort of program, and the easiest way of using it is to run the program from within Visual BASIC. With everything switched on and in position, operate the "DOWNLOAD" button to send the program to Lightbot via the infrared link. When the green light in the transmitter goes out, operate the "END" button to terminate the program. Lightbot, complete with new program, is then ready for testing. I could detect no difference in the performance of the program in its RCX code and the Sprit.OCX forms.

Improved tracking

The obvious way of avoiding the overshoot problem is to briefly reverse Lightbot when the line is detected. This backing up slightly also reduces the risk of wheel spin resulting in Lightbot going forwards beyond the line instead of turning sharply. The modified RCX code is shown in Figure 3.26. An obvious difference between this program and the original is that it only has program code on the dark side of the light sensor "brick". This is acceptable because Lightbot does one thing if the light value is at or below a certain level, and another if the value is above this level. It would not be possible if there was an overlap of the light and dark ranges, or a gap between them, but in most practical applications there is a straightforward light-to-dark transition.

The main program is the same as the original, and it is only the light sensor programming that is different. The main program sets Lightbot going forwards and this is the way things continue until the sensor detects the line. The five program "bricks" beneath the light sensor block are then performed, starting with both motors being set into reverse. A Wait instruction then keeps Lightbot in reverse for 0.2 seconds, and then motor C is set to the forward direction again. This turns Lightbot to the right, and another Wait instruction keeps it turning for 0.5 seconds. Both motors are then set to forwards operation again.

By this time the light sensor should be over a white section of the paper track again, and Lightbot will therefore continue to go forwards until the black line is encountered. The reverse, turn, and go forwards routine will then be repeated, keeping Lightbot inside the line. If the sensor should still be over the black line after one of these routines, further routines will be performed until the sensor is brought clear of the line. This method of control does not give particularly smooth movement around tight

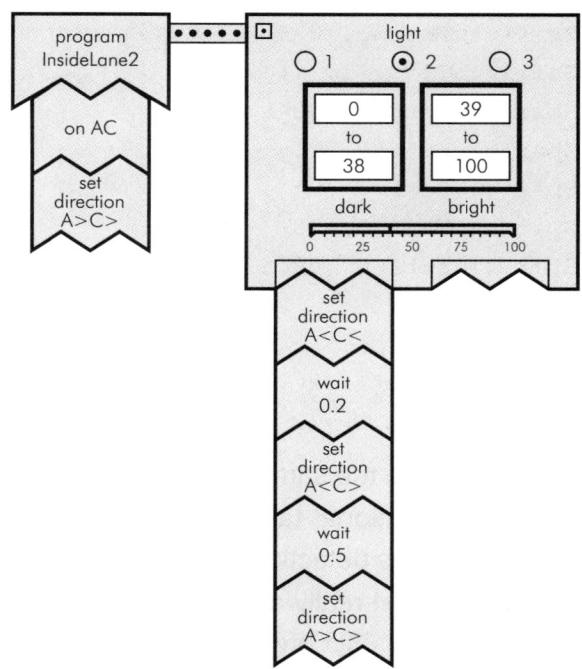

Fig.3.26 *The RCX code for improved tracking*

curves, but it does virtually guarantee that Lightbot will not cross the line and "do a runner".

VB version

The Visual BASIC version of the program uses the same form as the original, with the same code for the "END" button. This code replaces the original listing for the "PROGRAM" button.

```
Spirit1.InitComm
Spirit1.SelectPrgm 1
Spirit1.BeginOfTask 0
Spirit1.SetSensorMode 1, 4, 0
Spirit1.SetSensorType 1, 3
Spirit1.On "02" Spirit1.SetFwd "02" Spirit1.StartTask 1
Spirit1.EndOfTask
Spirit1.BeginOfTask 1
```

```
Spirit1.Loop 2, 0
Spirit1.If 9, 1, 1, 2, 41
Spirit1.SetRwd "02"
Spirit1.Wait 2, 20
Spirit1.SetFwd "0"
Spirit1.Wait 2, 50
Spirit1.SetFwd "02"
Spirit1.EndIf
Spirit1.EndLoop
Spirit1.EndOfTask
```

Task 0 is much the same as its equivalent in the original program, but it only calls one additional task. Like the RCX code version, this program only responds to the detection of darkness by the sensor. It is only fair to point out that there is not really any point in having a main task that calls up a single additional task. Simply merging the additional task into the main one would be functionally the same. I have arranged the program this way simply to make it a direct equivalent to the RCX code version.

The similarities between Task 1 and its RCX code equivalent should be self-evident. The If instruction performs a list of five commands if darkness is detected. As before, the motors are set in reverse and kept in that direction for 0.2 seconds by a Wait instruction. Then motor 0 is set to forward operation to turn Lightbot, and a second Wait instruction keeps it turning for another 0.5 seconds. Finally, both motors are set to forward operation. Again, I could detect no difference in performance between the Spirit.OCX and RCX code programs. The raw code downloaded to the RCX unit could well be the same regardless of which version is used.

Mechanical solutions

As mentioned previously, the slippage problem can be counteracted by mechanical means, and this is actually the better approach. Software can make Lightbot follow the line, but progress around the curves tends to be very slow and hesitant. A lot of time is spent going backwards. One way of getting improved results is to use some form of front wheel steering. The front wheel or wheels are steered using one motor, while the other is used to propel the robot. This approach is certainly possible using the Robot Invention System, which includes facilities for rack and pinion steering for example. Front wheel steering is certainly something that is interesting to try when you have some experience at building Lego robots, but it does not represent a good starting point.

Two simpler alternatives are to use front wheels that are pivoted but non-driven, or to use a tracked vehicle. The problem with the front wheels of Lightbot is that that must slip sideways over the track if sharp turns are to be accommodated. If the front wheels are allowed to pivot, they no longer have to skid sideways over the track, but can instead swing round to point in the appropriate direction and then move normally over the track. A tracked vehicle still has to do a certain amount of slipping sideways in order to accommodate tight turns, but the superior grip of the tracks is better at avoiding overshoot. It also seems to avoid the problem of one side driving the robot efficiently while the other simply skids over the track. Four-wheel drive would probably have a similar effect, but as the Robot Invention System includes a pair of tracks a tracked vehicle is the obvious choice.

Gripbot

The tracked version of Lightbot, which we will call Gripbot, is based on the Lightbot chassis. Figures 3.27 and 3.28 show two views of the completed robot. The only modification to the original chassis is to move the two black "rivets" near the front of the chassis one set of holes further forward. The holes previously occupied by the "rivets" will be used for the front axles. A slight modification is also required to the motor assembly. Instead of fitting the two mounting blocks at the rear of the motor assembly they are positioned at the front of it, and the assembly is mounted further back on the chassis (Figure 3.29). Both motors are fitted with 16-tooth gearwheels (Figure 3.30). The RCX unit can be fitted in its original position, but it is better to move it back one set of pegs/holes so that there is no gap between it and the motor assembly.

Fig.3.27 The Gripbot robot uses tracks to reduce slippage on paper circuits

Obviously the main change is that the original wheels are replaced with the tracks and the four white wheels that match them. However, things are not as simple as just replacing the original wheels with the new wheels and tracks. Each track tends to pull its pair of wheels together. Unless this effect is reduced to insignificant proportions there may be large amounts of slippage between the wheels and the tracks. Also, the tracks may tend to work free of the wheels. All that is needed to keep everything in the right place is a pair of extra 16 x 1 beams mounted slightly outboard of the existing 16 x 1 beams in the chassis. These extra beams can be seen in Figure 3.31, which also shows the way everything fits together. Two of the plates on the underside of the chassis have been removed so that the fixings for the axles can be seen. The parts used in one of the track assemblies are shown in Figure 3.32, which also helps to show how it all fits together. Note that all four axles are the same size for Gripbot, and are approximately 63 millimetres long.

Fig.3.28 Rear view of Gripbot showing the changes to the motor assembly

The front wheel assemblies are the more simple, and it is easier to fit these first. Push an axle through the appropriate hole in the chassis and secure it with a wide clip. Then add another wide clip over the axle on the other side of the chassis. Next a 16 x 1 beam is fitted over the axle, and another wide clip is added to keep the beam in

Fig.3.29 The modified motor unit

place. To complete the assembly the wheel is added, together with yet another wide clip to hold it in place. This process is then repeated on the other side of the chassis.

Start constructing one of the rear wheel assemblies by pushing an axle through the appropriate hole in

Fig.3.30 Gripbot uses a higher gear ratio

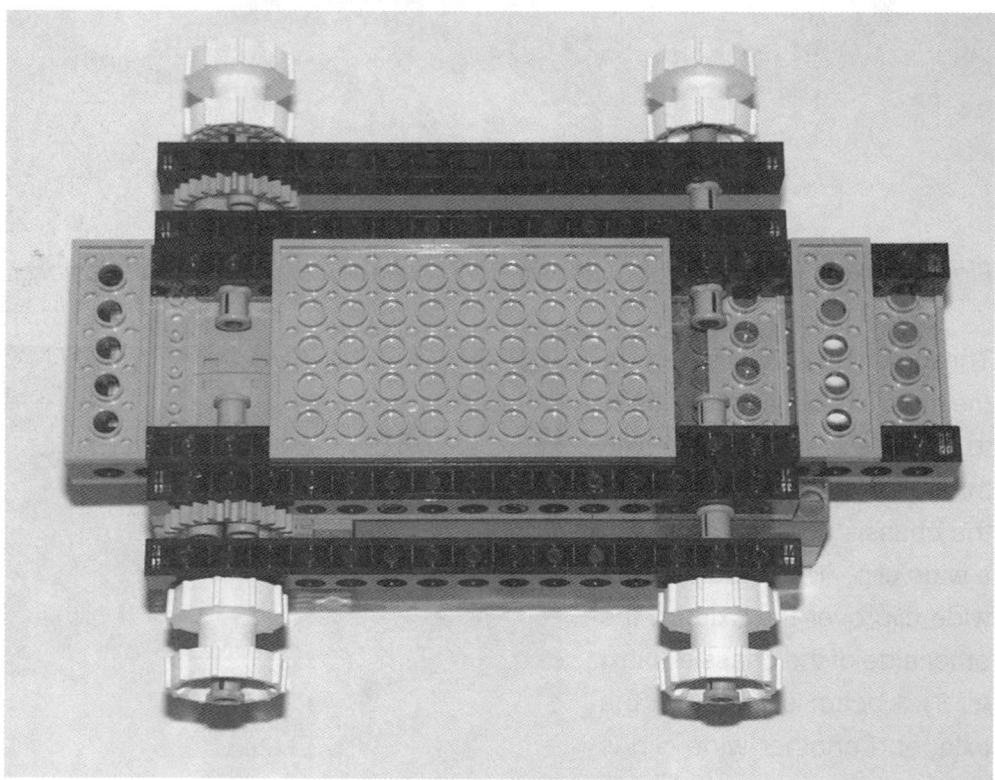

Fig.3.31 Extra beams are used to strengthen the track assemblies

the outer beam, and then through a 24-tooth gearwheel and the chassis. Adjust the axle so that it protrudes about seven millimetres through the chassis and then fit it with a wide clip. Next fit a 16-tooth gearwheel onto the axle followed by the wheel. The gearwheel may seem to serve no purpose, but it provides coupling from the axle to the wheel. Without it the axle will rotate but the wheel will remain stationary. To complete the assembly, add a wide clip to keep the wheel in position and then carefully fit the track onto the wheels. With this process repeated on the other side of the chassis and the bottom plates added, the chassis and track subassembly is finished.

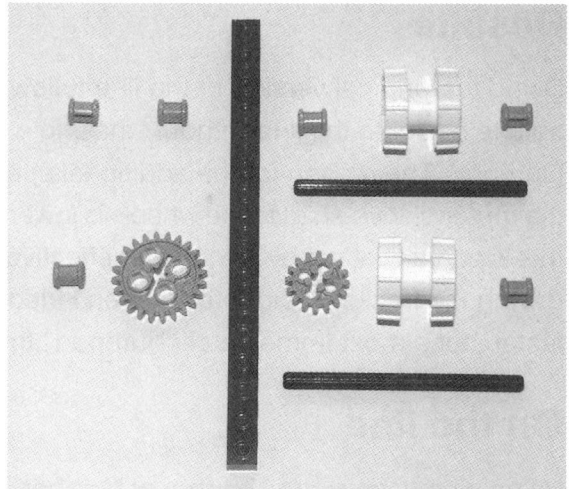

Fig.3.32 *The parts needed for one track assembly*

Fig.3.33 *The method used to mount the sensor on the front of the chassis*

The light sensor is fitted on the front of Gripbot using a right-angled 4 x 2 plate, but this time it does not fit onto the RCX unit. Instead a 6 x 2 plate is fitted onto the top of the chassis right at the front, and the light sensor is mounted on this, as in Figure 3.33.

Results

Using the original version of the line-following software, Gripbot proved well able to navigate the line without "making a break for it" and much faster than Lightbot. The reason for the additional speed is that the reduction ratio from the motor's drive-shaft to the wheels is lower at 4 to 6 instead of 1 to 5, although this is partially offset by the smaller effective wheel diameter and power lost in driving the tracks. Incidentally, when I tried 1 to 1 gearing Gripbot went even faster, but almost immediately "jumped" the line and made its getaway.

On the line

Using simple software Lightbot or Gripbot are able to stay within the line, but can they be made to follow the line? Ideally this type of thing requires two light sensors that are positioned on opposite sides of the robot. With the robot positioned such that the sensors are either side of the line, the controlling software can detect more than whether the robot is on or off the line. When the robot strays from the line, the sensors indicate which way it has deviated from the correct course. If the left-hand sensor detects darkness, the robot has strayed to the right, and if the right-hand sensor detects darkness the robot has wandered off course to the left. This makes it much easier to write effective control software for the robot, which can also be kept "on the straight and narrow" more efficiently.

Unfortunately, the Robot Invention System is only supplied with a single light sensor. Ultimately you may wish to obtain or build more sensors, but here we will make do with a single sensor. Using the oval track provided with the system, a single sensor is actually all that is needed. With the original line-following program the robot was moved straight forward when the sensor was over white paper, and turned to the right when the black line was detected. This caused the robot to stay within the black line and move around the track in a clockwise direction.

In order to make the robot follow the track in the same direction, but over the line, it is merely necessary to reverse the way the program operates. In other words, the robot must move forwards when it is over the black line, and turn to the right when it is not. The modified RCX code to provide this action is shown in Figure 3.34. This is the same as the original line-following routine, but the two "bricks" under the light sensor block have been swapped over.

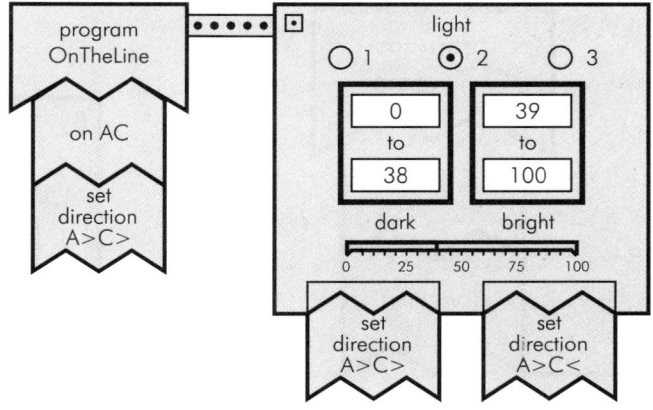

Fig.3.34 The modified code that keeps the robot on the line

This simple method works quite well when used with a manoeuvrable robot such as Gripbot, but it does have its limitations. The main problem is that it will only work properly if the robot stays on the line, or strays outside it. If the robot strays onto the white paper inside the line it will turn to the right and move away from the line. It may eventually pick up the line again, but in the meantime the robot will go "walkabout". It is for this reason the robot needs to be an agile type that will hug the line reliably. The robot should be started pointing in roughly the right direction and with the sensor on the line. If you start the robot in the middle of the track you will have to rename it Dizzybot, because it will simply go round and round in circles going nowhere.

Drunkbot

A robot having a single light sensor can be made to seek out and follow a line, even if that line does not curve in one direction only. However, the control software is inevitably somewhat more complex, especially if the robot must progress at a reasonable rate. Having tried various methods, the only simple one that worked reliably is shown in the RCX code program of Figure 3.35. In order to keep things manageable the program is shown in Figure 3.35 in two halves, but it is actually one long string of commands.

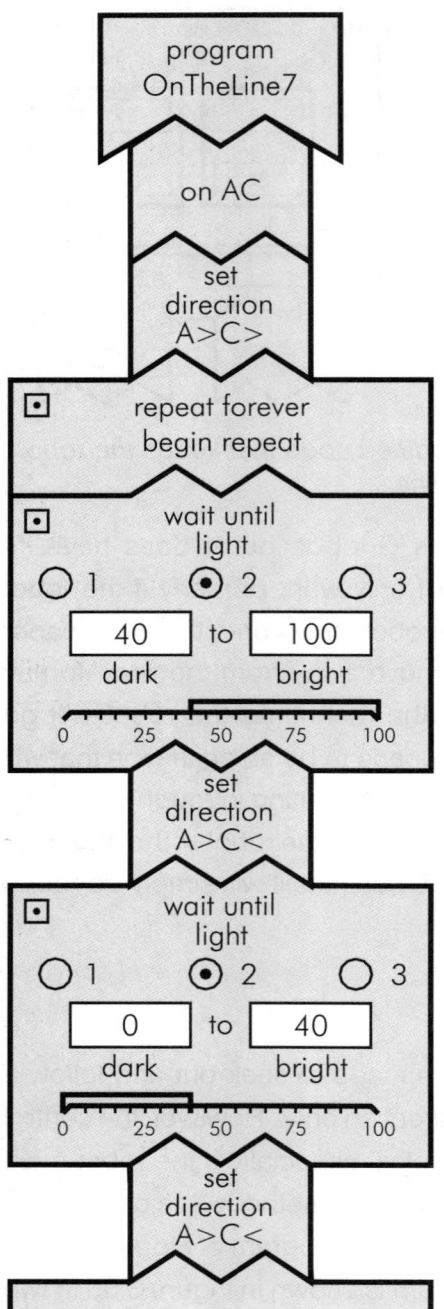

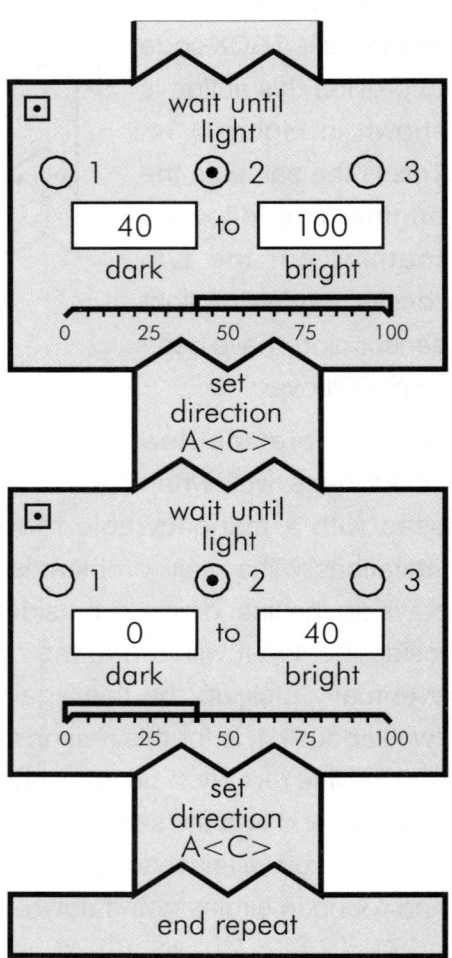

Fig.3.35 The RCX code for the true line-following program

The first two "bricks" simply initialise Drunkbot by switching on the motors in the forward direction. Things then move into the program proper, which is an infinite loop, or a "forever" loop in Lego MindStorms terminology. To be certain the program will operate correctly Drunkbot should be started with the sensor over the black line and it should be aimed slightly to the left. The Wait Until block keeps Drunkbot going forward until a high light level is detected, which means it has strayed off the line to the left. The direction of motor C is then changed so that Drunkbot turns to the right, and this continues until a low light level is detected again. The line has then been recaptured, but Drunkbot continues to turn to the right. The second Set Direction command is not really necessary, and I have simply included it to emphasise the point that Drunkbot continues to move to the right. A third Wait Until program block halts this action when a high light level is detected, which means that Drunkbot has strayed off the line on the right-hand side. The directions of the two motors are then reversed so that Drunkbot turns to the left. This continues until the line is recaptured again and a low light level has been detected, and then until a high light level is sensed as the line is lost again. By this time the program is back at the top of the loop again, and the whole process repeats indefinitely.

On the face of it Drunkbot will simply sway from side to side making no progress down the line. In reality there is significant forward progress as a robot of this type turns, and Drunkbot should gradually find its way around the track. This method seems to be very reliable, and once correctly into the weaving routine there seems to be no escape from the line. It is not particularly efficient though, and if asked to "walk" a straight line Drunkbot will still weave from side to side. It is from this that the polite version of its name is derived. In order to check that curves in either direction can be followed it is not necessary to make up your own test track. Simply check that Drunkbot can follow the track going clockwise or anticlockwise.

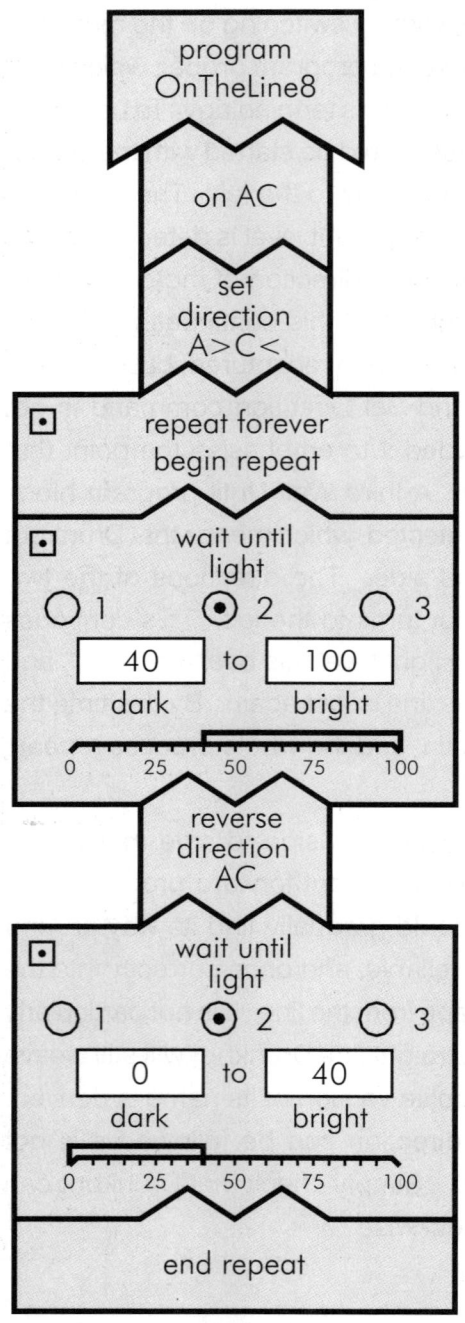

program
OnTheLine8

on AC

set
direction
A>C<

☑ repeat forever
begin repeat

☑ wait until
light

○ 1   ⊙ 2   ○ 3

| 40 | to | 100 |

dark   bright

0   25   50   75   100

reverse
direction
AC

☑ wait until
light

○ 1   ⊙ 2   ○ 3

| 0 | to | 40 |

dark   bright

0   25   50   75   100

end repeat

*Fig.3.36 The over-efficient line
following program*

Trancebot

The line following program can be somewhat simplified by using the Change Direction command. Instead of having one routine to turn to the left and another to turn to the right, the program can simply have one routine that reverses the direction of both motors (Figure 3.36). In general it is better to set motors to a specific direction instead of using an instruction that reverses the current direction. By setting a definite direction the action of the program is unambiguous, and it does not rely on the motor going in a certain direction when the command is issued. In this case the current direction of the motors is unimportant, and we simply wish to reverse whatever direction of rotation that happened to be, so the Reverse Direction command is perfectly adequate.

In practice this "improved" version of the program seems to be a little too efficient, and it is likely to produce the robot equivalent of a trance! There will be side to side movement with no significant progress along the track. This highlights a problem when writing software for many forms of specialist hardware, especially robots. Getting a program to perform properly is not just a matter of getting the program flow correct. What seems like a perfectly plausible program may fail due shortcomings or unexpected behaviour from the robot. Having written a program it is a matter of trying it out to see what happens, and looking carefully to see exactly what is going wrong if the robot does not perform as expected. If things go wrong it is still necessary to check programs for errors, but do not be surprised if a little "fine tuning" is required with programs that are fine "on paper".

In this case, the behaviour of the "improved" program gives a possible way of making Drunkbot perform in a more sober fashion so that progress around the track is faster. Try preceding each Set Direction command with another command of the same type that sets both motors to forward operation. Figure 3.37 shows part of the program that has been modified in this way. The program now has numerous instructions that seemingly perform no worthwhile function, but if you try this less efficient program you will find that Drunkbot actually moves around the track much more quickly. The instructions that set both motors to forward operation nudge Drunkbot along the track but do not produce sufficient forward movement to prevent the line tracking from operating properly. On the face of it, these extra instructions will have no effect as they will be almost instantly overridden by the Set

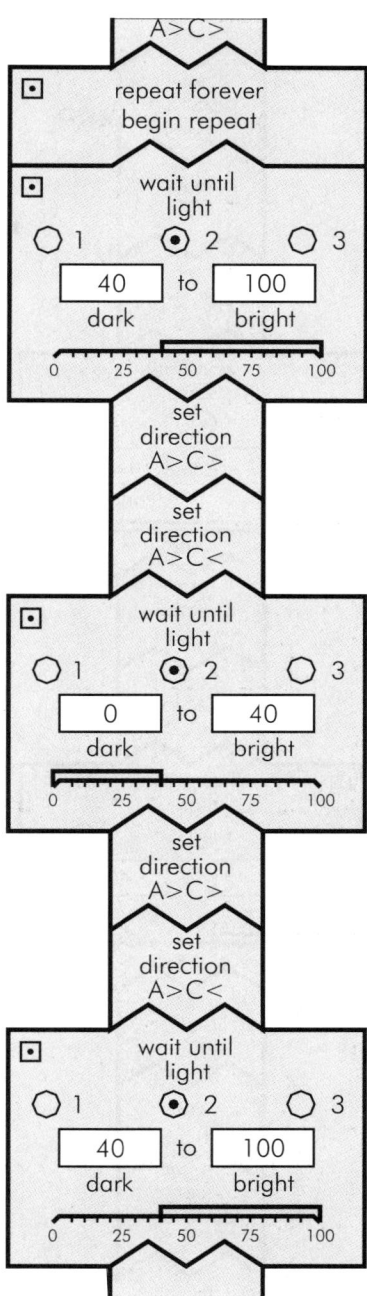

Fig.3.37 *The modification to give faster forward movement*

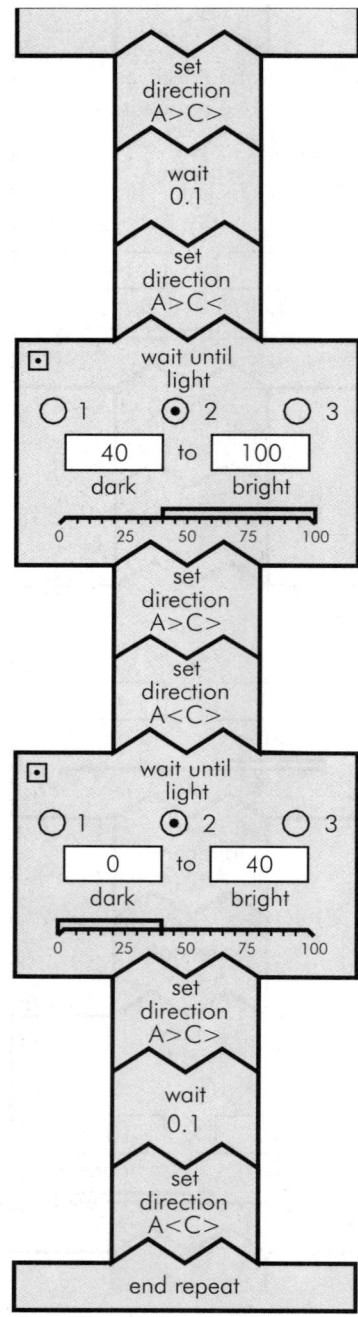

Fig.3.38 This modification further improves speed

Direction commands that follow them, but in practice RCX code does not seem to perform fast enough for this to happen.

Things can be taken a step further by adding a couple of Wait instructions to prolong the forward movement, and move Drunkbot faster along the track. The modified end section of the program is shown in Figure 3.38. This seems to work quite well in practice, but only if the minimum duration of 0.1 seconds is used in the Wait command. Anything more than this produces a tendency for tracking to be lost, with Drunkbot frequently swivelling around and heading off down the track in the opposite direction.

Between the lines

It would probably be possible to devise ways of making the line-following robots operate more efficiently, but with only one sensor and one line there will inevitably be a fair amount of hunting for the line. As pointed out previously, things are much easier if the robot is equipped with two sensors. The controlling software is then able to determine whether the robot has strayed "off the beaten track" to the left or the right, and apply corrective measures. There is another means of achieving much the same effect, and it only requires a single sensor. It is just a matter of using two lines and one sensor instead of two sensors and one line.

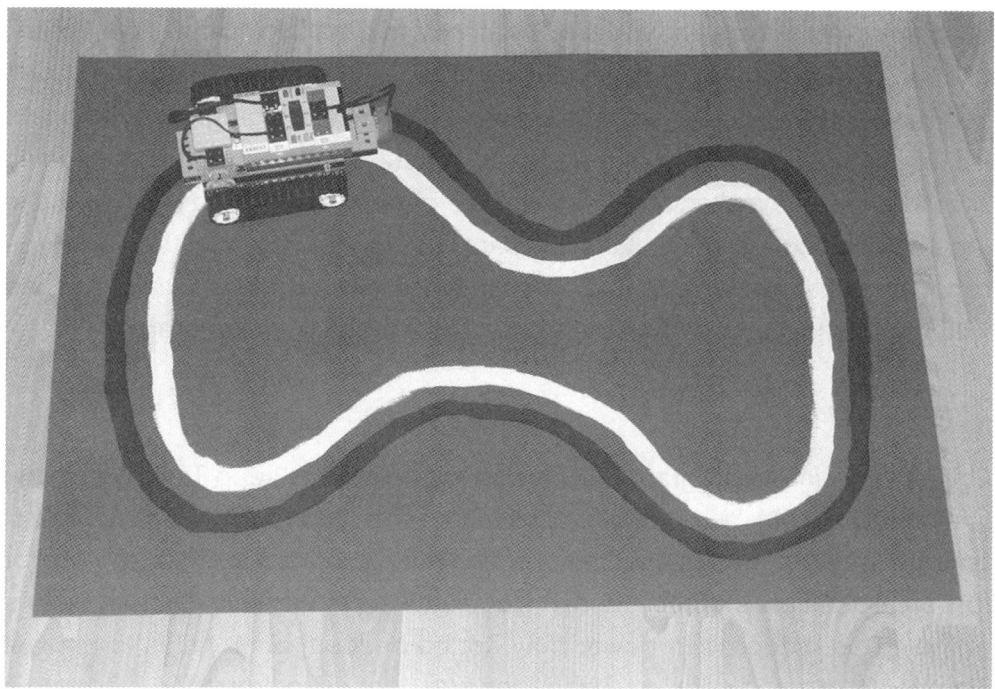

Fig.3.39 The prototype two-line track in operation

There is more than one way of arranging things, but in this example we will use a track that is based on paper that has a mid-tone. As in the Lego track there is a black line, but inside this there is a white line. The prototype track was made using a piece of heavyweight paper intended for pastel and watercolour paintings, but a cheaper and lighter paper should do the job just as well. The colour of the paper is unimportant, but it must provide in-between light readings from the sensor. Use the robot to take readings from the light and dark areas of the Lego track, and then test some likely looking papers to see if they produce readings that are reasonably well spaced from the light and dark readings. The piece of paper must be fairly large, and A1 or a near imperial equivalent represents the minimum that is likely to give satisfactory results.

Try to produce a reasonably challenging course, but avoid any really tight curves. Figure 3.39 shows the prototype track with a robot in action. The lines should be about 15 millimetres apart and at least 15 millimetres wide. I used artists' acrylic paints, but virtually any white paint should do. If necessary, use two or three coats of the white paint to give good contrast with the paper. You have to be choosier with the black paint, since a high gloss type could fail to give low readings from the sensor. A matt-black paint is preferable and should provide excellent contrast with the paper. There is no need to make the track particularly neat. If the lines are literally a bit rough at the edges the robot should still be able to track them properly, and when you have finished with the track it can be framed and used as a piece of modern art!

Trainbot

Only a very simple program in RCX code is needed to keep Trackbot on the lines (Figure 3.40). No modifications are needed to the robot, which we will rename Trainbot for this application. The main program sets the motors for forward operation, switches them on, and sets the power level for both motors at five. This reduction in power may not be needed, but it helps to prevent Trainbot from being derailed. If the lines are widened to around 25 millimetres it should not be necessary to use reduced power. Two light sensor blocks are used because the program must detect three light ranges and perform a different task for each range. With a mid-range reading of between 36 and 44 the sensor is between the lines, and both motors are set for forward operation. With a low reading of 35 or less the sensor is over the black line and Trainbot must turn to the right. Accordingly, motors A and C are respectively set for forwards and reverse operation. At readings of 45 or more the sensor is over the white line and Trainbot must turn to the left. Motor A is therefore set in reverse and motor C is set for forward operation.

Of course, the threshold levels mentioned here are simply provided as examples. The actual levels used if you follow these examples may need to be changed to suit the particular sensor and track you are using. The program of Figure 3.40 is only suitable for clockwise rotation of the track. For counter-clockwise operation the light and dark commands should be swapped, as in Figure 3.41.

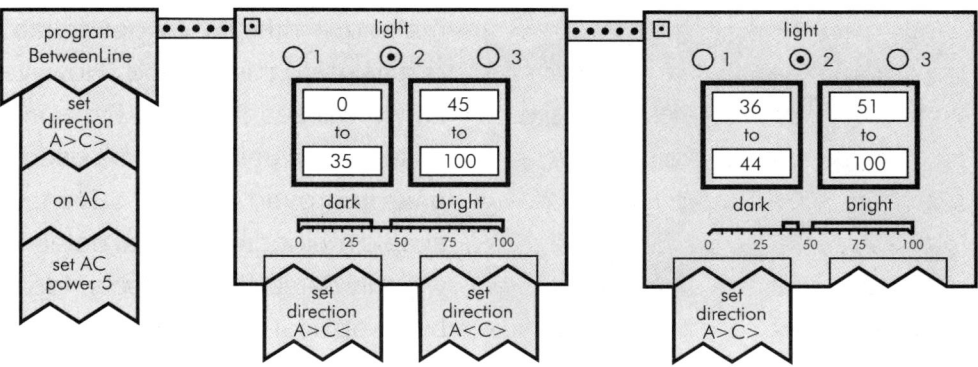

Fig.3.40 The RCX code for tracking between two lines

Object detection

The Robotics Invention System is supplied with two touch sensors that can be used to detect that an explorer style robot has collided with an object, so that it can manoeuvre its way free and continue its exploration. This is a rather clumsy way of doing things though, and the light sensor offers an interesting alternative. Rather than waiting for the robot to hit obstacles, the light sensor can try to detect them so that collisions can be avoided. The change in light level as an obstacle is approached is unlikely to be very great,

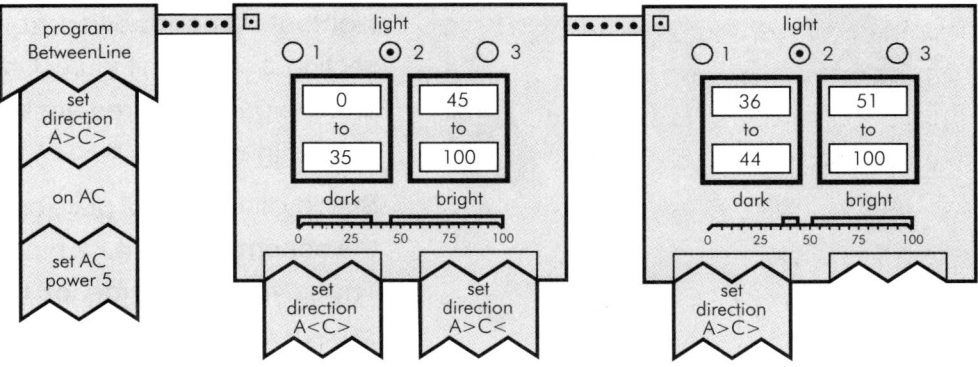

Fig.3.41 The counter-clockwise version of the program

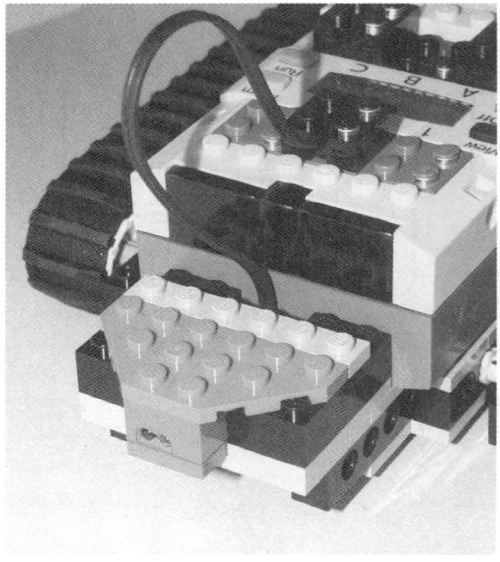

Fig.3.42 The light sensor assembly

making it difficult to get reliable results using this method. However, as we will see later, the RCX unit's infrared transmitter can be used to give improved results. Even so, simple methods of optical detection will only detect large objects. Ideally both optical and touch sensors would be used. This is a subject that we will investigate in chapter 4, and for the time being we will only consider the implementation of light sensors.

Lightbot and Gripbot are both suitable as the basis for these experiments, but in this case the light sensor must "look" forward rather than down at the ground. It is not difficult to devise suitable methods of fitting the light sensor, but it will tend to get knocked out of position unless the mounting is well designed. The method described here is suitable for use with Gripbot,

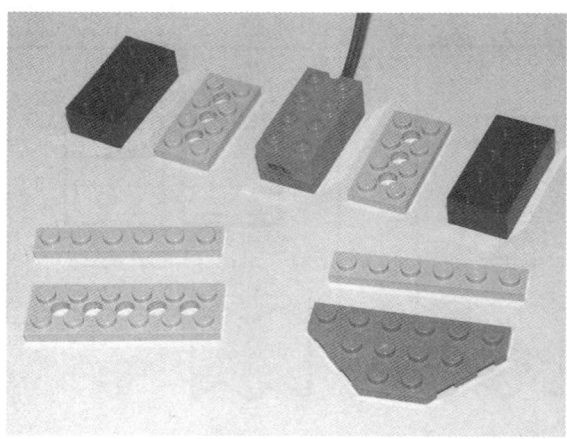

Fig.3.43 The parts for the sensor assembly

but it should not be difficult to devise something similar for Lightbot. The modified front section is shown in Figure 3.42, and the parts required are laid out in Figure 3.43.

Start by fitting a 4 x 2 plate onto the bottom of each 4 x 2 block. These fit either side of the sensor, but one row of holes/pegs further back. A 6 x 2 plate and a 6 x 1 plate fitted on the underside hold the three main pieces together, as shown in Figure 3.44. The green plate and the second 6 x 1 plate are then added to the top of the assembly, which is then mounted at the front of Gripbot. The finished robot is shown in Figure 3.45.

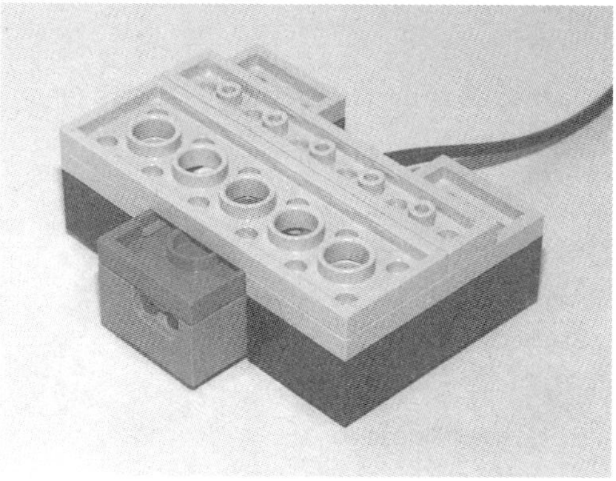

Fig.3.44 *Fitting the plates on the underside of the assembly*

Proxbot1

We are using the light sensor as the basis of a proximity detector, and it is from this that the Proxbot name is derived. In this first version the detection process relies on the fact that the light readings tend to increase as the sensor gets close to a large object. This happens because the light sensor includes a light emitting diode (LED), and this sheds more light on an object as it gets closer to that object. This is the full Visual BASIC listing for Proxbot1:

Fig.3.45 *The finished robot complete with new front section*

```
Private Sub Command1_Click()
Const Reading = 1

With Spirit1
.InitComm
.SelectPrgm 3

.BeginOfTask 0
.On "02"
.SetSensorType 1, 3

.SetSensorMode 1, 0, 0
.SetFwd "02"
.Loop 2, 0
.SetVar Reading, 9, 1
.SubVar Reading, 2, 12
.Wait 2, 20
.If 9, 1, 1, 0, Reading
.SetRwd "02"
.Wait 2, 100
.SetFwd "02"
.Off "0"
.Wait 2, 150
.On "0"
.EndIf
.EndLoop
.EndOfTask
End With
End Sub

Private Sub Command2_Click()
Spirit1.CloseComm
End
End Sub
```

Like the previous Visual BASIC programs, this one requires the Spirit.OCX control and two buttons to be included on the form. One button runs the main program that downloads the code to the RCX unit, and the other terminates the program once the transfer is complete.

Variables

This program makes use of a variable, and variables operate in a very simple way when using Spirit.OCX. With normal programming a variable is a name that is used to store a value that will change during the time that the program operates. In this case we wish to store readings from the light sensor so that they can be subjected to some simple arithmetic that will modify the returned readings. The values could therefore be stored in a variable called Reading. In reality a variable is a piece of data stored in a memory location, but it is easier for the programmer to refer to the variable by a meaningful name rather than by its memory address, which is simply a multi-digit number. The programming language makes life easier for you by accepting variable names in programs, which the language converts into the correct address when the program is compiled.

With Spirit.OCX variables operate in a much more basic fashion. With a computer such as a PC there are large amounts of random access memory (RAM), which is used to take the program as well as being used for variables and other temporary data storage. The microcontroller in the RCX unit has a very different arrangement, with programs stored in a form of read only memory (ROM). This type of memory has the advantage of not suffering from total amnesia when the RCX unit is switched off, and your programs remain ready for use next time the unit is switched on. The microcontroller has some additional registers that can be used for temporary data storage, and these are effectively its RAM. There are 32 of these registers that are numbered from 0 to 31, and Spirit.OCX only accepts these numbers, not names such as Reading or TotalValue. Each variable contains a 16-bit signed integer, which in decimal terms means a whole number in the range –32768 to +32767.

Although Spirit.OCX is not very accommodating when it comes to variable names, Visual BASIC is very flexible in this respect. You can declare a variable in Visual BASIC, and it will then substitute the appropriate value anywhere that variable appears in your program. This feature is used here to enable the variable called Reading to be used in the program, with visual BASIC actually using the value 1 when this variable is specified. In other words, values read from the light sensor are stored in the variable called Reading, which is actually register 1 in the RCX unit. This is not really necessary in this case since you do not have to be a genius to remember that you are using one variable called 1, but it does demonstrate the way in which this system operates. If you use several registers for data storage it can be very difficult to keep track of everything unless this method of variable name substitution is adopted.

Program operation

The initial part of the program assigns a value of 1 to the variable called Reading, which is strictly speaking a constant, since the value of Reading is always 1. It is the value stored in register 1 that changes and is a true variable. Communication with the RCX unit is then initialised and program 3 (4) is selected. The first and only task is then commenced. Initially both motors are set to move Proxbot forwards, and then the sensor type and mode are respectively set to light and raw. In raw mode the value read from the sensor is in the range 0 to 1023, which gives better resolution than the 0 to 100 of the percent mode. In practice there are often problems with various types of noise on the readings, and the improvement in usable resolution might be much less than the figures given above would suggest. Raw mode is still a better choice though, as it should give somewhat greater flexibility.

The rest of the task is an infinite loop. First the light sensor is read and the returned value is placed in the variable called Reading, which is really register 1 of the RCX unit. A value of 12 is then deducted from Reading, the program waits 200 milliseconds (0.2 seconds), and a new light reading is taken by the If instruction. The If instruction compares the new reading with a value equal to 12 less than the previous reading. If the new reading is the lower value, it is likely that Proxbot is approaching an object, and the following code is performed. This may seem to be the wrong way round, but in Raw mode increased light produces reduced readings. Hence the program is designed to search for a reduction in readings, not an increase. If no decrease is detected, the program goes back to the beginning of the loop and repeats the measuring and testing process. If the following code is performed, Proxbot goes into reverse for one second, turns to the left for 1.5 seconds, and then goes forward again. This enables Proxbot to go exploring without hitting anything substantial, apart from the odd glancing blow. These minor accidents occur because the sensor is only sensitive to objects that are directly in front of it. As Proxbot is fairly wide, one of the tracks might therefore clip something that is just out of the sensor's "vision".

It is possible to vary the sensitivity of the proximity detector by altering the value deducted from the value placed in Reading. A value of more than 12 reduces the sensitivity, and a lower value increases it. In practice a value of much less than 12 will probably give erratic operation with Proxbot doing the robot equivalent of seeing things. It is worth experimenting with different values to see if performance can be improved. The prototype detected most objects at a range of 100 millimetres or more, but some were virtually rammed before they were detected.

Proxbot2

The performance of the proximity detector can be improved by using the RCX unit's infrared communicator to transmit some "light" for the sensor to detect. Like most semiconductor photocells, the one used in the RCX unit seems to readily detect infrared radiation close to the visible red part of the spectrum. In fact most semiconductor photocells offer peak sensitivity to "light" at these wavelengths. This is the listing for Proxbot2:

```
Private Sub Command1_Click()
Const Reading = 1
```

```
With Spirit1
.InitComm
.SelectPrgm 3

.BeginOfTask 0
.On "02"
.StartTask 1
.SetSensorType 1, 3
.SetSensorMode 1, 0, 0
.SetFwd "02"
.Loop 2, 0
.SetVar Reading, 9, 1
.SumVar Reading, 2, 90
.If 9, 1, 0, 0, Reading
.SetRwd "02"
.Wait 2, 100
.SetFwd "02"
.Off "0"
.Wait 2, 150
.On "0"
.EndIf
.EndLoop
.EndOfTask

.BeginOfTask 1
.Loop 2, 0
.SendPBMessage 2, 0
.Wait 2, 2
.EndLoop
.EndOfTask
End With

End Sub
```

```
Private Sub Command2_Click()
Spirit1.CloseComm
End
End Sub
```

One obvious change from the original program is the inclusion of a second task. This loops indefinitely, transmitting a dummy message at 20 millisecond intervals. The main task is very similar to the original, but there are a couple of differences. The If instruction checks to see if the new reading is greater than the previous one, not less than, and a value of 90 is added to the previous reading rather than being deducted from it. This may seem to be round the wrong way, but it demonstrates the point that in this case we are looking for changes from one reading to the next. If you modify the If instruction to check for a large decrease in value, and deduct 90 from the first reading instead of adding it, the program will still work.

The important point to keep in mind here is that the infrared transmitter does not work continuously, but instead fires short bursts of "light". Some readings will be taken when the transmitter is operating, others will be taken while it is switched off. This gives the required changes from one reading to the next. It may seem a bit "hit and miss" to check for an increase or a decrease in value instead of testing for both. However, the pulses are sent at a rate of 50 per second and the readings are taken in rapid succession. This ensures the rapid detection of an object when it comes within range. Proxbot2 is better in this respect than Proxbot1, which takes 200 milliseconds or so to take and check pairs of readings.

A value of 90 is added to the previous reading before it is compared to the current reading. On the face of it such a high value will give poor sensitivity, but Proxbot2 seems to be very much more efficient than Proxbot1. Even using a value as high as 90, sensitivity is much improved and the risk of spurious detection is greatly reduced. Any reasonably substantial object is detected at a range of around 300 millimetres or more. Even a small object such as an AA size battery is detected at a range of around 25 to 60 millimetres provided it is directly in front of the sensor.

Showbizbot

Showbizbot is mechanically the same as Proxbot1, and the only differences are in the software. The Showbizbot name is appropriate because this robot seeks out the bright lights. Its basic action is to turn until it detects a light, and then it heads for that light. This is the full listing for Showbizbot:

```
Private Sub Command1_Click()
Const Reading = 1

With Spirit1
.InitComm
.SelectPrgm 2

.BeginOfTask 0
.On "02"
.SetSensorType 1, 3
.SetSensorMode 1, 0, 0
.SetRwd "0"
.SetFwd "2"
.Loop 2, 0
.SetVar Reading, 9, 1
.SubVar Reading, 2, 40
.Wait 2, 5
.If 9, 1, 1, 0, Reading
.AlterDir "02"
.Wait 2, 10
.SetFwd "02"
.Wait 2, 800
.Off "02"
.EndIf
.EndLoop
.EndOfTask
End With
```

```
End Sub

Private Sub Command2_Click()
Spirit1.CloseComm
End
End Sub
```

This program has strong similarities with the Proxbot1 software, and it is derived from that program. Like the original, it is designed to detect a drop in the light reading from the sensor. Initially Showbizbot turns in a counter-clockwise direction rather than going forwards and it does this to search for a light source. As the sensor swings round towards the light it will produce increasing readings, but as it swings past it and starts to turn away, the readings will reduce. The commands following the If instruction are then performed, and the first of these alter the directions of both motors for 100 milliseconds. The point of this is to compensate for the fact that Showbizbot has gone beyond its quarry. This slight backtracking should adjust its aim so that it is pointing at the light source. The motors are then set to go forwards, and Showbizbot sets off towards the light source. The end section of the routine switches off both motors after eight seconds, which should be long enough for the light source to be found. Obviously a longer delay can be set here if necessary.

Practically any torch placed on the ground and aimed towards Showbizbot should give the desired result, but in practice the marksmanship of this robot will probably not be perfect. Results are better with some torches than others, and variations in the strength of the beam are a problem with many torches. These variations can cause falls in the readings while the sensor is still turning towards the light source. If this happens, the aim can be improved by having Showbizbot continue to turn in a counter-clockwise direction for a short while rather than turning back in a clockwise direction.

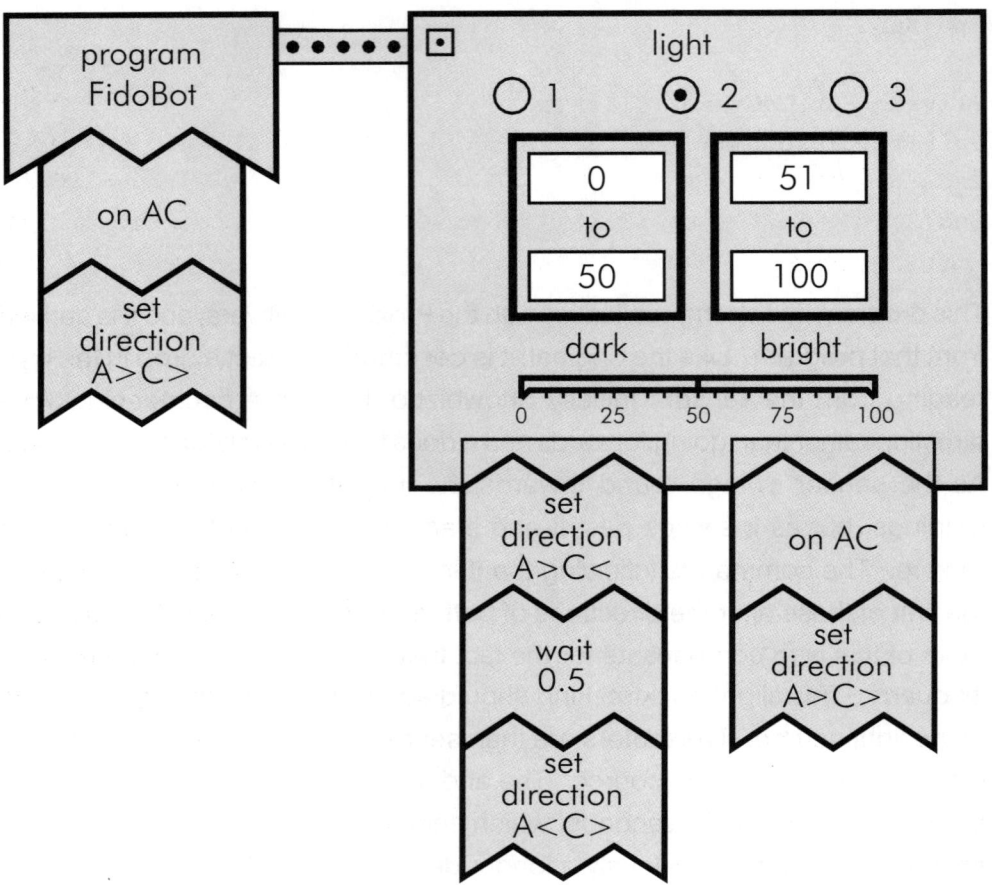

Fig.3.46 The RCX code for Fidobot

Fidobot

The idea of this robot is that it follows on a lead like a dog, but in this case the lead is actually a torch beam. Aim the torch at Fidobot and it moves forward. Aim the torch slightly to the right and it moves to the right, aim it to the left and Fidobot moves to the left. In both cases, once back in the torch beam Fidobot goes forward again. Figure 3.46 shows the RCX code for Fidobot.

The main program simply switches on both motors and sets them to the forward direction. The same short routine is used on the light side of the light sensor block, and Fidobot therefore goes forward if the light sensor provides a high enough reading. I found that results were quite good using the default threshold level of 50/51, but this can be changed if necessary. There is a slight problem when producing the code for the dark side of the light sensor. This must turn Fidobot in the appropriate direction, but there is no way of telling which direction that is. Therefore, the direction of motor C is reversed for 0.6 seconds, turning Fidobot to its right. If that does not locate the beam from the torch, the directions of both motors are changed, turning Fidobot to its left until the beam is located.

3   Lightbots

Touchbots

Skid free

The touch sensors supplied with the Robotics Invention System are simply switches that are normally open, but close when the small yellow buttons are fully depressed. In normal electronics terminology this type of component is known as a micro-switch. Although less sophisticated than the light sensor, the touch sensors nevertheless have numerous uses and can be utilized in robots of various types. The first robot featured in this chapter is another

Fig.4.1 Touchbot has a pivoted front wheel that permits tight turning

Fig.4.2 The "antennas" at the front of Touchbot are actually bumpers

rover style robot, and like the previous robots it is steered by varying the directions of the drive wheels. Unlike the previous robots, it is a three-wheeled vehicle where the third wheel is free to rotate (Figures 4.1 and 4.2).

Arranging this third wheel is a little more complex than one might think. On the face of it, there is no problem if the pivot point is directly above the wheel, or above a point directly between the two wheels that are used here to provide better stability. In practice things do not work too well with the pivot directly above the wheel, because the wheel tends to stay in whichever direction it happens to be pointing instead of swivelling round to match the direction in which the robot travels. The front wheel can tend to act like a skid for much of the time. The easiest way around this is to have a short arm, with the wheel fitted at one end and the pivot at the other. This provides some leverage that helps to swivel the wheel around so that it is aimed in the appropriate direction.

For this method to work properly the wheel assembly must be free to rotate through 360 degrees. This tends to make the front of the chassis higher than the rear even if the drive wheels at the rear are somewhat larger than the pivoted wheel at the front. This is tolerable, but things are neater if the chassis has the stepped construction used here. This keeps everything more or less level and permits the front wheel assembly to rotate freely.

Obviously a single touch sensor mounted at the front of the unit could be used to detect collisions so that the robot can manoeuvre away from obstructions. However, the button on each touch sensor is very small, which gives it limited efficiency when used unaided. The robot could ram things left, right, and centre, but the sensor would only detect things that were rammed dead centre. Using two sensors plus the bumpers at the front of the robot ensure that hitting practically anything will activate one or other of the sensors. Having two sensors enables the software to determine which side of the robot has hit something, so that it can manoeuvre out of trouble more efficiently. Of course, the bumpers help to avoid any damage occurring to the robot when it hits something, and also help to avoid damage to anything the robot hits.

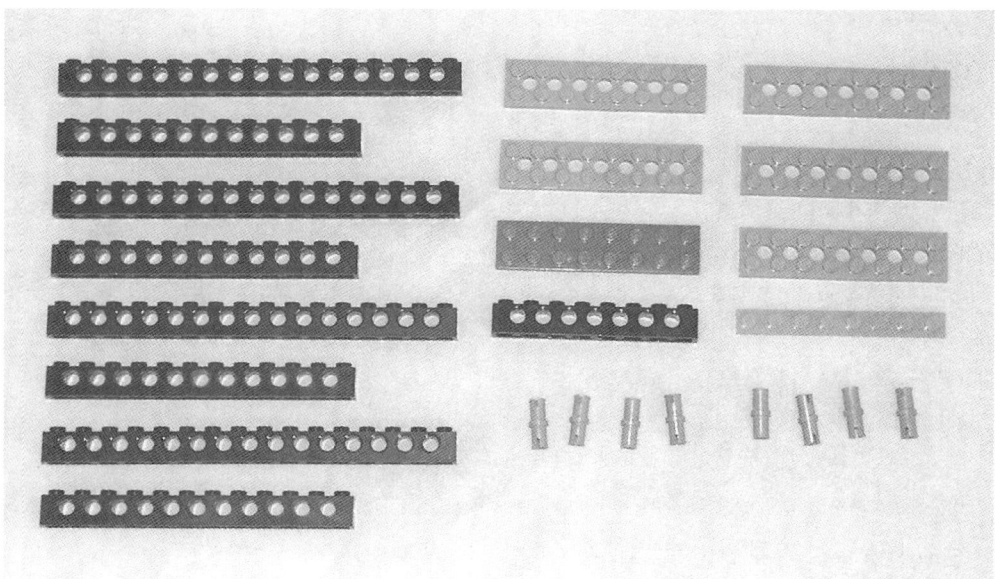

Fig.4.3 The parts required to build the main chassis

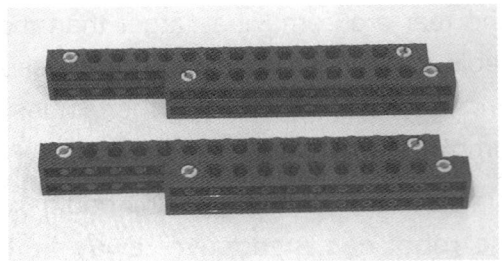

Fig.4.4 The completed sub-assemblies

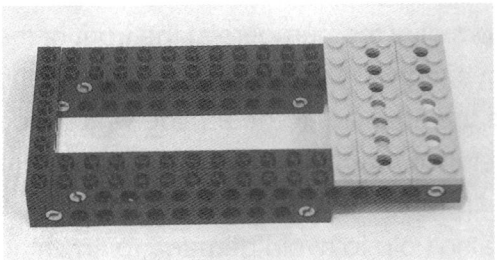

Fig.4.5 Joining the two sides of the chassis

Step 1 (Figures 4.3 - 4.6)

Figure 4.3 shows the parts required for the main chassis. Start by making one of the side pieces by fitting a 16 x 1 beam on top of a 12 x 1 beam. Then repeat this process with the remaining three sets of beams so that you have four identical sub-assemblies. Then fit pairs of the sub-assemblies together using four "rivets" per pair, to give the two completed side pieces (Figure 4.4). Note the deliberate misalignment of the 12 x 1 beams with the 16 x 1 beams. Next the two sides are joined together by adding

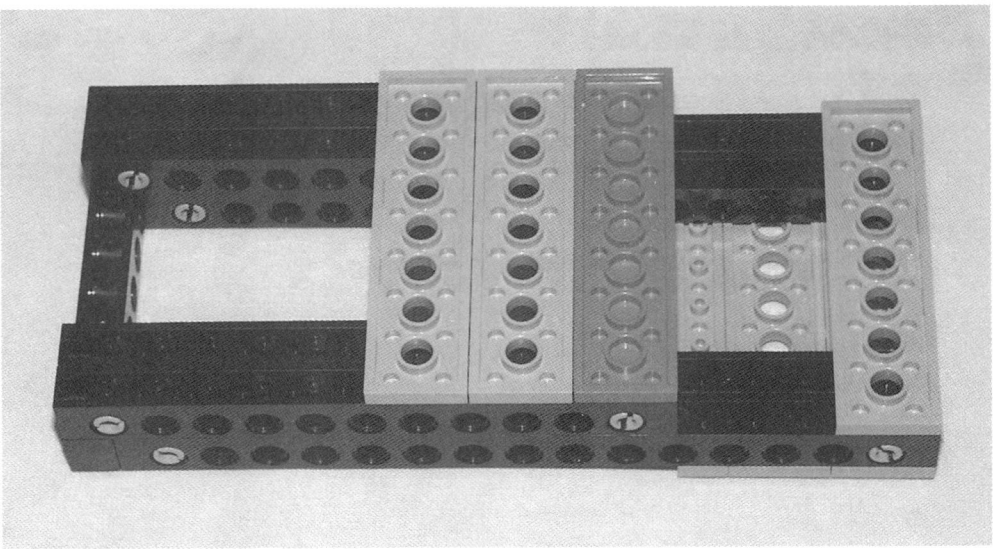

Fig.4.6 The underside of the completed chassis

an 8 x 1 beam at the rear, followed by two 8 x 2 and one 8 x 1 plate at the front (Figure 4.5). To complete the main chassis, turn it over and add the remaining plates on the underside (Figure 4.6).

Step 2 (Figures 4.7 - 4.10)

The parts required for the rear section of the chassis are shown in Figure 4.7. Fit a 10 x 1 beam on top of another beam of the same size, and then add a

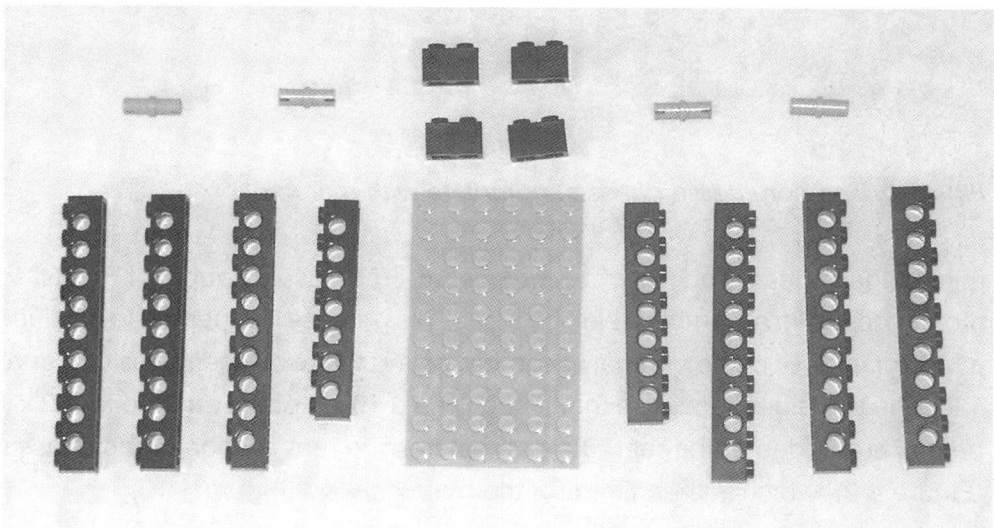

Fig.4.7 The components needed to build the lower chassis

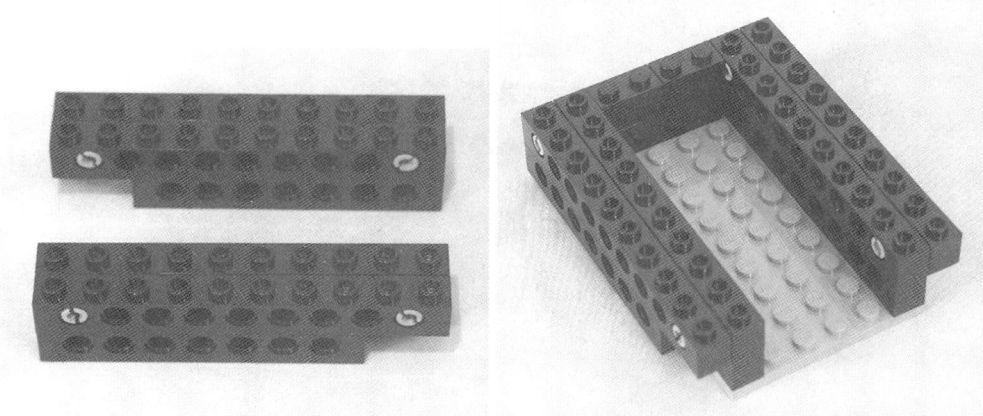

Fig.4.8 The complete side pieces   *Figs.4.9 The finished lower chassis*

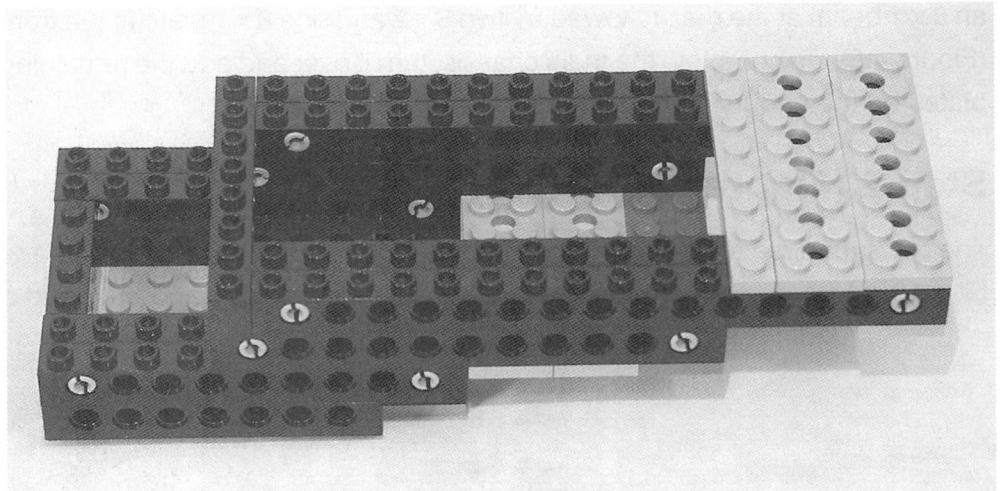

Fig.4.10 The completed chassis, complete with rear section

third of these beams on top of an 8 x 1 beam. Then use a couple of "rivets" to mount these two assemblies side by side. This process is repeated to produce a "mirror image" of the original assembly, giving the two assemblies of Figure 4.8. The two side pieces are then fitted onto a 10 x 6 plate, and the four 2 x 1 beams are added at the rear. This completes the rear section of the chassis (Figure 4.9), which is then fitted on the main chassis (Figure 4.10).

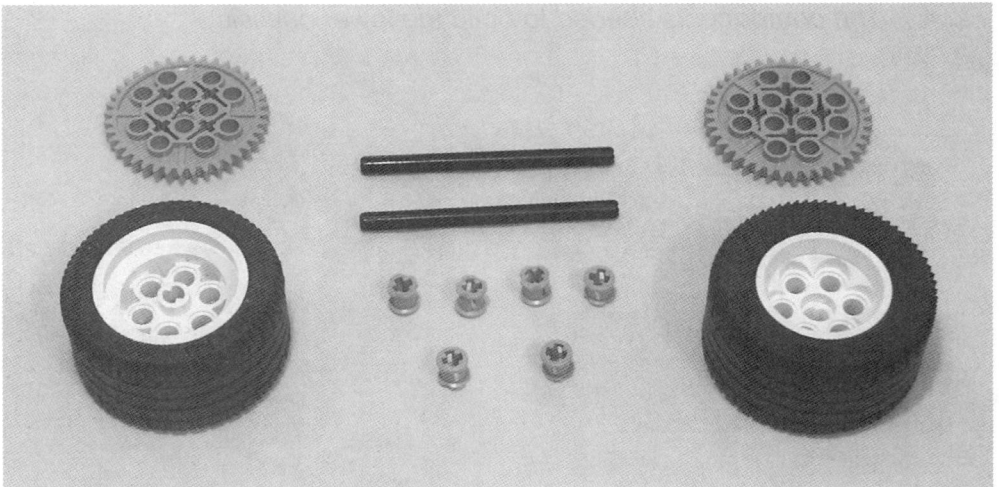

Fig.4.11 The parts required for the rear wheel assemblies

Step 3 (Figures 4.11 and 4.12)

The rear wheels and associated parts are shown in Figure 4.11. The wheels are the wide type about 50 millimetres in diameter including the tyres. As the left and right wheel assemblies are identical I will only describe the construction of one. Start by placing a long fixing "nut" on the end of an axle about 63 millimetres long. Then fit a wheel onto the axle, and as always, the more concave side should be facing outwards (towards the fixing "nut" already on the axle). Push the wheel as far onto the axle as possible, followed by another long fixing "nut". A 40-tooth gearwheel is then added, again pushing it as far onto the axle as possible. The axle is then fitted onto the bottom beam at the rear of the chassis, in the second hole in from the rear. Another long fixing "nut" is then added to hold the wheel assembly in place. With both wheels added the rear section should look like Figure 4.12.

Fig.4.12 The rear wheels fitted to the chassis

Step 4 (Figures 4.13 - 4.15)

The motor assembly is quite simple and uses just eight parts, which includes the two motors (Figure 4.13). The two 8 x 1 plates go on the undersides of the motors and fix them together. The two yellow 4 x 2 plates are fitted on top of the motors, and also help to hold them together. To complete the motor assembly the 16-tooth gearwheels are fitted onto the spindles of the motors (Figure 4.14). The finished assembly fits onto the rear section of the chassis (Figure 4.15).

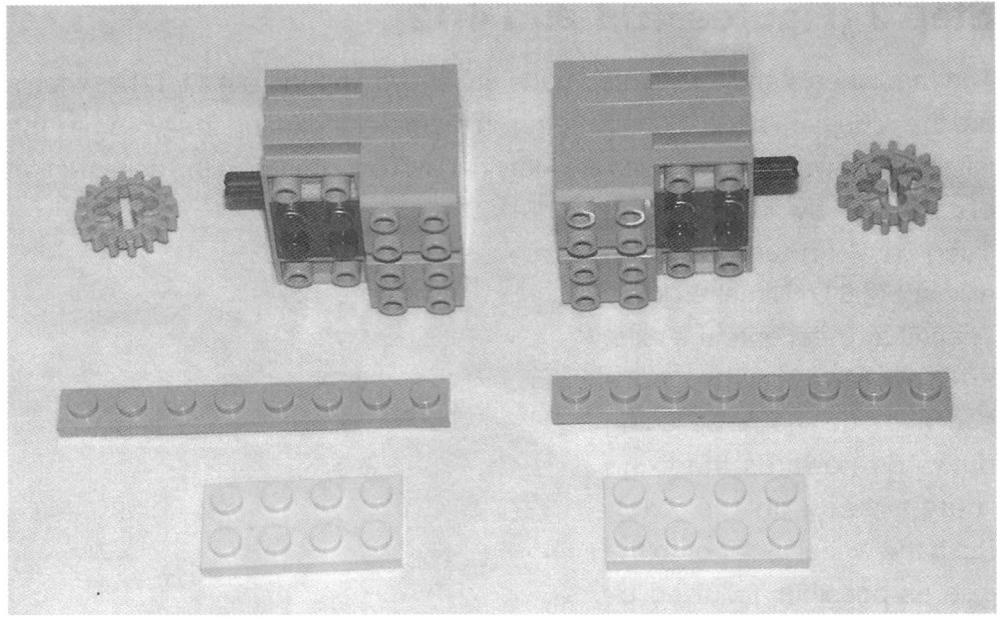

Fig.4.13 The parts required to build the motor assembly and mount it on the rear of the chassis

Fig.4.14 The finished motor assembly

Fig.4.15 The motor assembly installed on the rear of the chassis

Fig.4.16 The parts required for the front wheel assembly

Step 5 (Figures 4.16 - 4.20)

The chassis is now ready to take the front wheel assembly. There are two front wheels, but they share a common axle and are mounted close together so that they effectively form one wide wheel. The parts required for this assembly are shown in Figure 4.16. The wheels are the small type of about 30 millimetres in diameter including the tyres. They are placed on a short axle (about

Fig.4.17 The front wheel assembly

Fig.4.18 The complete pivot arm

Fig.4.19 The fully finished assembly

30 millimetres in length), either side of the blue T-shaped piece. Two "nuts" are used at the ends of the axles to help keep everything together (Figure 4.17). Leave everything very slightly loose so that the wheels will turn freely. Next make another sub-assembly using the two short grey coloured beams, another 30-millimetre axle, a small right angle piece, and two long fixing "nuts" (Figure 4.18). The two sub-assemblies are then joined together using a third 30-millimetre axle and two long fixing nuts (Figure 4.19). Finally, the whole assembly is mounted on the chassis using a 46-millimetre axle and two long fixing "nuts". A short small pulley is used

Fig.4.20 The front wheel assembly fitted onto the front of chassis

between the chassis and the wheel assembly to act as a washer (Figure 4.20). Provided everything is not pressed together too tightly, this will ensure that the wheel assembly can rotate reasonably freely.

Step 6 (Figure 4.21)

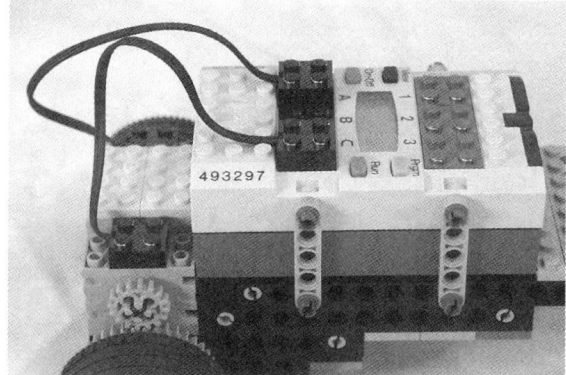

Fig.4.21 *The RCX unit mounted on the chassis. The side pieces are optional*

The next step is to mount the RCX unit on the chassis immediately in front of the motor assembly, and to connect it to the motors (Figure 4.21). The side pieces that secure the RCX unit to the chassis are optional. They help to strengthen the assembly, but also make it more difficult to change the batteries.

Step 7 (Figures 4.22 - 4.24)

The bumper and touch sensors are mounted on a raised platform at the front of the unit. The parts required are shown in Figure 4.22. Start by mounting

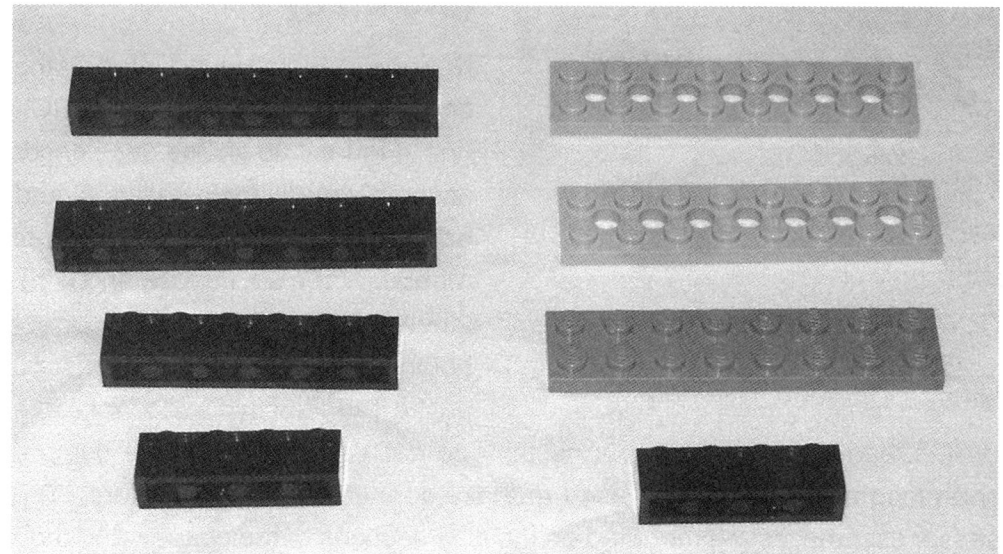

Fig.4.22 *The components needed to make the raised platform*

Fig.4.23 The start of the platform   *Fig.4.24 The completed platform, ready to take the sensors*

the two 8 x 1 beams, the 6 x 1 beam, and the two 4 x 1 beams on the front of the chassis (Figure 4.23). Next add two 8 x 2 plates at the front of the platform, one above and one below, and add a third 8 x 2 plate at the rear of the platform, on the top side (Figure 4.24).

Fig.4.25 The parts required for one sensor assembly, plus a finished assembly

Step 8 (Figures 4.25 and 4.26)

Now the sensors are mounted on the chassis and connected to the RCX unit. Figure 4.25 shows one sensor assembly ready for installation, and one set of parts. Each sensor is mounted on a 2 x 2 block and a 1 x 2 beam, which raise it to the correct height. Note that the touch sensors, unlike the light type, do not have built-in leads. Instead, the connections to each sensor are made using a short lead of the same type used to make connections to the motors. The sensor assemblies are mounted on the side sections of the chassis and then connected to inputs "1" and "3" of the RCX unit (Figure 4.26).

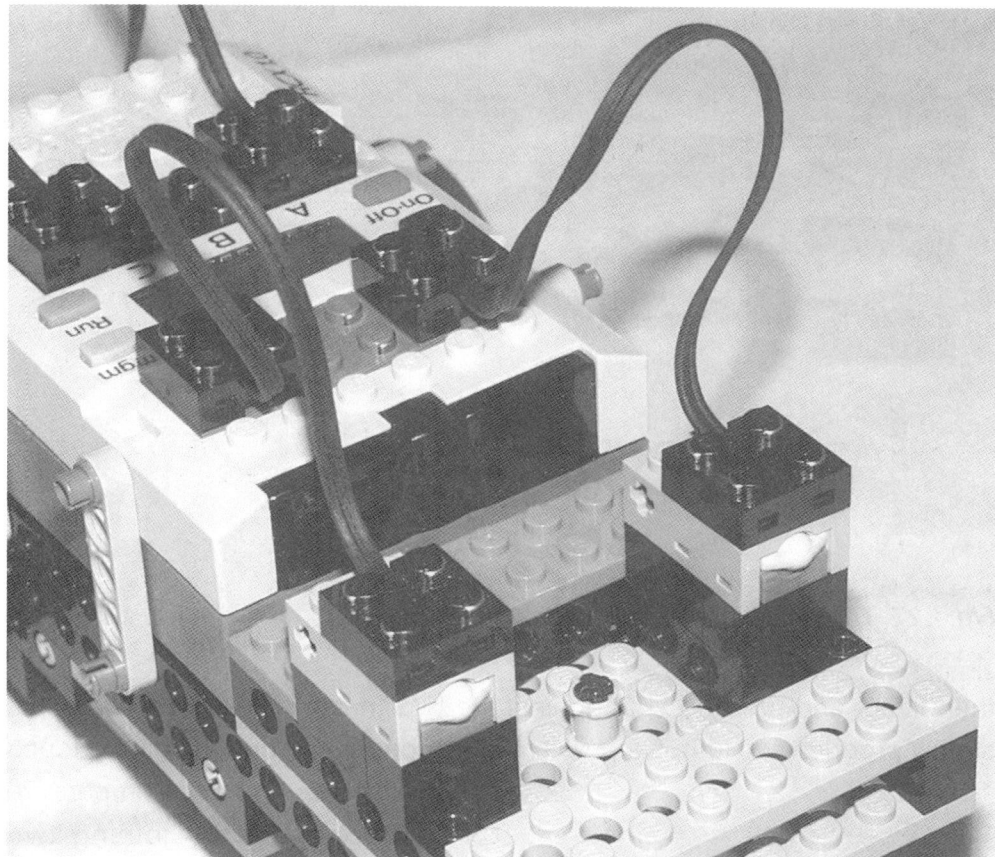

Fig.4.26 The sensors in position on the raised platform

Step 9 (Figures 4.27 and 4.28)

The final stage is to build and fit the bumpers, making sure that they operate the two touch sensors correctly. Figure 4.27 shows the parts required, and a partially assembled bumper assembly appears in Figure 4.28. Start by fitting the two longer (about 79 millimetres long) axles into the double-end blue coloured piece having a hole through the middle. This hole will eventually take the shorter axle, and the whole assembly will pivot on this axle. Next fit two wide "nuts", one on each axle, so that they fit against the blue coloured piece. The two pieces that are designed to connect axles at right angles are

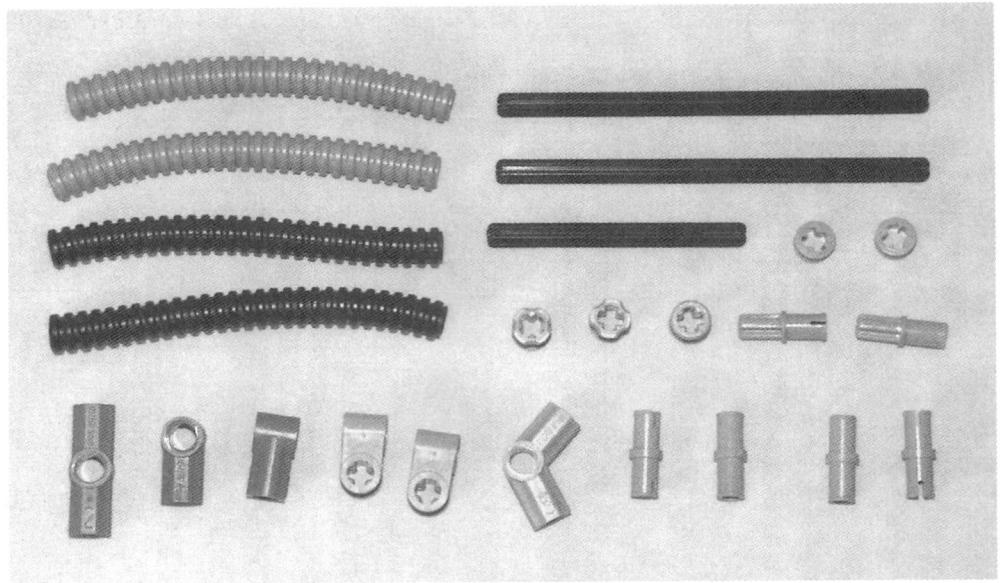

Fig.4.27 The parts required for the two bumpers

then fitted, and pressed tight against the two "nuts". These last two items simply produce two lumps that will reliably operate the buttons on the touch sensors. The two bumpers are each made by joining two flexible tubes (about 72 millimetres long) using the grey connecting pieces that are the same at both ends. Use tubes of whatever colours you like - they are mechanically the same. The bumpers are fixed to the blue pieces at the ends of the axles using two more of these connectors, and then they are joined to the grey piece that has its two sections forming an angle of about 135

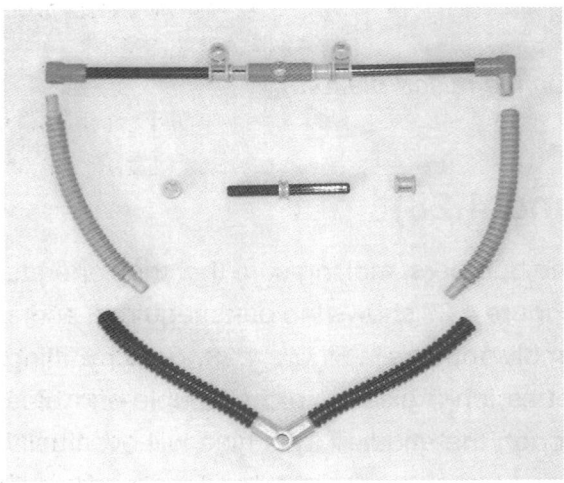

Fig.4.28 The partially constructed bumpers

degrees. Note that the two connectors needed to do this are different to the ones used elsewhere in the assembly. One end has a standard fitting, but the other end is like a short piece of axle, and it is this end that fits into the angle piece.

Step 10 (Figures 4.29 and 4.30)

To complete the robot the bumper is mounted on the two middle holes of the plates at the front of the robot (Figure 4.29). Thread the shorter (about 46 millimetre) axle through these holes, and place a fixing "nut" on the bottom end of the axle. Then add a long "nut" on the top of the axle and push it as far onto the axle as possible. The bumper is then added to the axle, fitting the blue piece at the rear first. Then the front of the assembly is folded backwards so that the grey angled piece can also be fitted onto the axle. Finally, a "nut" is added at the top of the axle to hold everything in place. With everything assembled correctly you should end up with the finished robot of Figure 4.30. Check that the bumpers operate the touch sensors correctly.

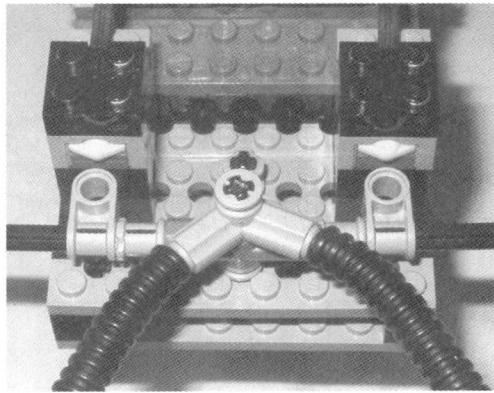

Fig.4.29 The bumpers mounted on the raised platform

RCX Software

The RCX code for Touchbot is shown in Figure 4.31. The main part of the program simply turns on both motors and sets them going forwards. One branch of the program monitors the sensor connected to input 1 of the RCX unit, and responds to this switch being turned on. This happens when the left side of Touchbot's bumper collides with something. The direction of both motors is then set as reverse, and a Wait instruction keeps Touchbot in reverse for one second. Motor A is then set to go forwards, which turns Touchbot to

Fig.4.30 The finished robot, ready for testing

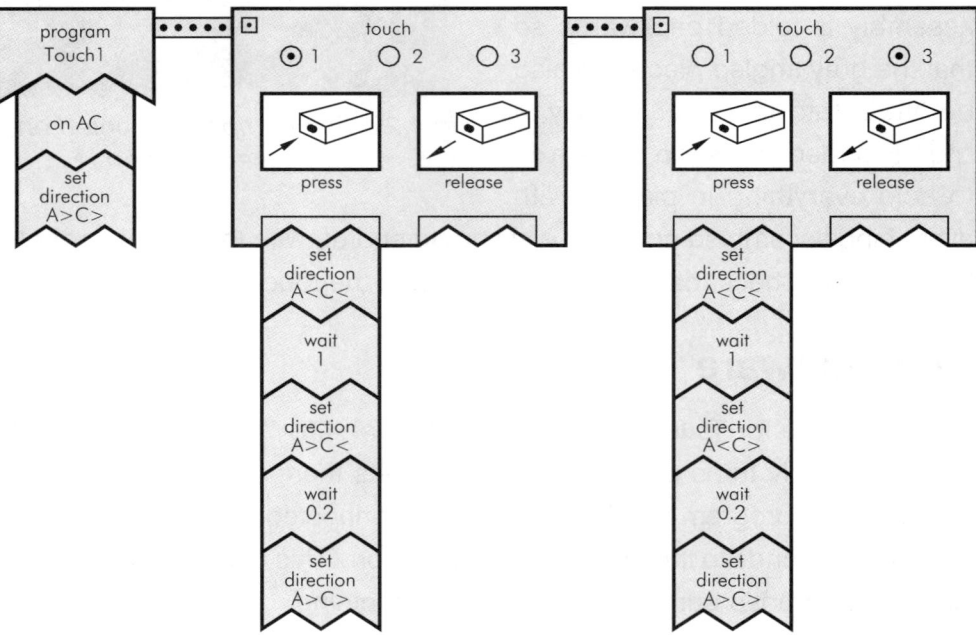

Fig.4.31 The RCX code for Touchbot

the right. A Wait instruction maintains this action for just 0.2 seconds, but due to the speed and manoeuvrability of Touchbot this is sufficient for it to turn through about 90 degrees. Both motors are then set to go forwards, and Touchbot goes merrily on its way once again. The software routine for the sensor connected to input 3 is much the same, but after reversing the robot turns to the left instead of the right.

There is a slight problem when downloading programs to the Touchbot RCX unit, because the two touch sensors and the bumpers tend to get in the way. They by no means fully block the windows on the RCX unit though, and reliable contact via the infrared link should be established if the transmitter unit is simply positioned a little higher than normal.

VB Software

This Visual BASIC program provides the same action as the RCX code. Like the programs in the previous sections, it requires a form that carries the Spirit.OCX control, a button marked "DOWNLOAD" (Command1), and another button marked "EXIT" (Command2).

```
Private Sub Command1_Click()
With Spirit1
.InitComm
.SelectPrgm 3
.BeginOfTask 0
.On "02"
.SetFwd "02"
.SetSensorType 0, 0
.SetSensorMode 0, 0, 0
.SetSensorType 2, 0
.SetSensorMode 2, 0, 0
.StartTask 1
.StartTask 2
.EndOfTask

.BeginOfTask 1
.Loop 2, 0
.If 9, 0, 1, 2, 500
.SetRwd "02"
.Wait 2, 100
```

```
.SetFwd "0"
.Wait 2, 20
.SetFwd "02"
.EndIf
.EndLoop
.EndOfTask

.BeginOfTask 2
.Loop 2, 0
.If 9, 2, 1, 2, 500
.SetRwd "02"
.Wait 2, 100
.SetFwd "2"
.Wait 2, 20
.SetFwd "02"
.EndIf
.EndLoop
.EndOfTask
End With
End Sub

Private Sub Command2_Click()
Spirit1.CloseComm
End
End Sub
```

The first subroutine in the listing is really the main program, and it is applied to the "DOWNLOAD" button. After some setting up the program starts by switching on both motors and setting them to move Touchbot forwards. Some further setting up then sets the type and mode for both sensors. With touch sensors you can use them in touch mode, or as here, simply select "none" as the type and read them in "raw" mode. This gives a reading of 1023 when the switch is open and a much lower value of around 37 when it is closed. Finally, the main program starts Task 1 and Task 2, and then terminates.

Tasks 1 and 2 monitor the two touch sensors, and are essentially the same. If we consider Task 1 first, the Loop command turns this routine into an indefinite loop. The If command reads the sensor on input 1, and the five values set these parameters:

9   Read a sensor

0   Read sensor 0 (input 1)

1   If the reading is less than…

2   the constant value…

500   500, then do this…

In other words, when the sensor on input 1 is activated, the following lines of code are performed. These lines set the motors into reverse for one second, set motor A to go forwards for 0.2 seconds, and then set both motors to go forwards. There is a direct correlation here between each line of Visual BASIC code and each block in the original program in RCX code.

Task 2 is much the same, but in the If command it is sensor 2 (input 3) that is read. Also, it is output 2 (motor C) that is set to go forwards for 0.2 seconds. Again, there is a direct correlation between this code and original RCX version.

The second subroutine is applied to the "EXIT" button, and this just terminates communications with the infrared transmitter and closes the program.

Switch mode

As already pointed out, the Visual BASIC version of the program uses the sensors in Raw mode, but they can be used in switch mode if preferred. It is just a matter of changing the four lines that set up the sensors, plus the two If statements that read them. These are the modified lines for setting up the sensors:

```
.SetSensorType 0, 1
.SetSensorMode 0, 1, 0
.SetSensorType 2, 1
.SetSensorMode 2, 1, 0
```

Previously the sensor type was set to "none" for both sensors, but in the modified code a value of 1 rather than 0 is used to set the sensor type. This sets both inputs to operate with switch type sensors. Instead of setting Raw mode using a value of 0, the mode is set to Boolean using a value of 1 in each of the SensorMode commands. This gives simple True/False readings, or returned values of either 0 or 1 in other words. Under standby conditions the returned value is 0, but it changes to 1 when a switch is activated.

These are the modified If statements. They are respectively the modified lines for Task 1 and Task 2.

```
.If 9, 0, 2, 2, 1
.If 9, 2, 2, 2, 1
```

The constant at the end of each statement has been changed from 500 to 1, and the middle value has been changed from 1 to 2. This means that the following program lines are performed if the value returned from the sensor is 1, and a value of 1 is returned when the switch is activated. The modified program therefore performs in exactly the same way as the original. If you load the original version of the program into the RCX unit and view the values returned from one of the sensors, they should be something like 1023 and 35 to 40, as explained previously. If you try the same thing with the modified program the returned values should only be 0 and 1.

Ditherbot

There is a potential snag with all the rover style robots described so far, and it centres on the fact that they always turn by the same amount when an obstruction is detected. This can result in the robot getting stuck in corners, where it simply goes from side to side. In most cases the robot will eventually work its way out of the corner, but this could take a long time, and is not guaranteed. Making the robot turn through a greater angle will reduce the problem. It is then unlikely to get stuck in a corner, but it can still end up exploring in a fixed pattern so that it is just going over the same ground time and time again.

The way around this problem is to make the robot less fixed in its ways. A more dithery robot will actually explore its surroundings more thoroughly than one that follows a fixed procedure with military precision. What is needed is a random element in the software to make the robot turn by a different and unpredictable amount each time a collision occurs. As we saw in chapter 2, the Wait instruction can be used to produce a delay of random duration, and on the face of it this is well suited to our present needs. In practice there is a slight drawback in that only the maximum delay can be controlled. Ideally it would be possible to set both the maximum and minimum durations. This would avoid having the occasional very brief turn that would result in the robot going almost straight back where it came from. One way around the problem is to use a variable to produce your own random numbers, which can then be in any desired range within the limits of the Wait command. Whether this method gives a truly random number is debatable, but purely academic. It gives the desired effect, which is all that really matters.

Ditherbot software

This is the Visual BASIC listing for a program that uses a variable to produce a randomised turn time in the range 0.05 to 0.4 seconds. Despite its randomness this version of Touchbot, which we will call Ditherbot, is actually more efficient.

```
Private Sub Command1_Click()
With Spirit1
.InitComm
.SelectPrgm 2
.BeginOfTask 0
.On "02"
.SetFwd "02"
.SetSensorType 0, 1
.SetSensorMode 0, 1, 0
.SetSensorType 2, 1
.SetSensorMode 2, 1, 0
.SetVar 1, 2, 5
.StartTask 1
.StartTask 2
.EndOfTask
.BeginOfTask 1
.Loop 2, 0
.If 9, 0, 2, 2, 1
.SetRwd "02"
```

```
.Wait 2, 100
.SetFwd "0"
.Wait 0, 1
.SetFwd "02"
.EndIf
.SumVar 1, 2, 1
.If 0, 1, 0, 2, 40
.SetVar 1, 2, 5
.EndIf
.EndLoop
.EndOfTask

.BeginOfTask 2
.Loop 2, 0
.If 9, 2, 2, 2, 1
.SetRwd "02"
.Wait 2, 100
.SetFwd "2"
.Wait 0, 1
.SetFwd "02"
.EndIf
.EndLoop
.EndOfTask
End With
End Sub

Private Sub Command2_Click()
Spirit1.CloseComm
End
End Sub
```

The basic structure of the program is much the same as before, but there are a few changes and additions. The method used to generate the random number is to have a counter that counts from 5 to 40, is then reset to 5, counts to 40 again, and so on. The value in this counter is used to set the time for which the robot turns after it has hit an obstruction, and a range of 5 to 40 equates to turn times of 50 to 400 milliseconds. Obviously some means of incrementing the counter is required, and it is simply incremented by one each time sensor 1 is read. I do not know how quickly each loop is performed, but it probably takes no more than a second or two for the counter to be cycled through its full range of values. Ideally the counter would cycle very much faster than this, and with a relatively slow rate of change it is theoretically possible for Ditherbot to get stuck in a corner, with the same value occurring each time the counter is read. In reality this is virtually impossible, and there was certainly no sign of any problem of this type when testing Ditherbot.

There is only one change in the main program (Task 0), and this is the addition of a SetVar instruction to initialise the variable that acts as the counter. The first value in this instruction selects the variable to be altered, and the next value sets the source for the new value. In this case we are using variable 1, and it is set at the constant specified in the third of the three values, or 5 in other words. The counter is incremented by the SumVar instruction in Task 1. This adds the specified value to the variable. The first number in this instruction indicates which variable must be incremented, and the next value indicates the source for the value to be added. In this case a value of 2 means that the third number in the instruction must be added, and this number is of course one.

So far we only have a routine that increments the counter on each loop of Task 1. We also need the counter to be reset when it goes past a value of 400. This is achieved using an additional If instruction in Task 1, and the five values in the If statement have this effect.

0   If the value in variable...

1   number 1

0   is greater than…

2   the constant value...

40   40, then do this…

Normally this condition will not be met, but when the count reaches 41 the following SetVar instruction is performed, and the counter is then set back to 5.

Having generated the pseudo-random number it must be used in the two Wait instructions that control the period that Ditherbot turns. This merely entails changing the two values in the existing Wait instructions. The first number is 0 to indicate that a value held in a variable must be used, and the second value is 1, which is the variable that contains the value.

When this version of the software is run the difference to the original should be readily apparent. Instead of always turning through about 90 degrees when an obstruction is hit, Ditherbot should turn through various angles from around 25 degrees to about 180 degrees. More importantly, there should be no tendency to get stuck in corners any more.

If you wish to try the Wait command's built-in random facility, use the original version of the program with this line to replace each of the Wait instructions that control the turn time of the robot.

```
Spirit1.Wait 4, 40
```

This gives a delay in the range 0.01 seconds to 0.4 seconds. It would presumably be possible to mathematically manipulate the random numbers to bring them within a desired range. For example, in this case we require a range of 5 to 40. A random number from 1 to 36 could be generated and placed in a variable. Adding 4 to the variable would then give a random number from 5 to 40. As a simple programming exercise, you might like to try this method of controlling Ditherbot.

Ballet

Ditherbot can be made more spectacular if it is made to spin round a few times before setting off in a new direction to continue exploring the terrain. It is likely to be somewhat less efficient at exploring because it will occasionally resume its original course and hit the obstruction once again. On other occasions it will go back in the direction it just came from, and will explore previously travelled territory. In order to make Ditherbot do pirouettes and turn it into Balletbot it is merely necessary to boost the number range of the random number generator. This increases the time for which the robot spins before setting off the explore "pastures new". A range of about 150 to 500 is about right, and this represents spin times of approximately 1.5 to five seconds. In Task 0 the line that initialises variable 1 must set its starting value to 150. This is the modified program line:

```
.SetVar 1, 2, 150
```

The second If structure in Task 1 must also be altered, as shown here:

```
.If 0, 1, 0, 2, 500
.SetVar 1, 2, 150
.EndIf
```

With reasonably fresh batteries this robot can spin quite rapidly, as will soon become apparent if you try this version of the software. You might like to experiment with other software for this robot. For example, using routines similar to those used to follow a line on the floor, it should be possible to get the robot to go around the edge of a room. There is plenty of scope for experiment here.

Executive toy

If you are fed up with various types of robot vehicle and would like to try something a bit different, this executive toy might "fill the bill". It is a form of

Fig.4.32 The finished executive toy, ready to make a big decision

gadget known as a decision-maker, and they are alternatively known as "heads or tails" machines. The basic idea is to have a gadget that helps stressed executives to make the right decision when something important comes up, such as tea or coffee, and one lump of sugar of two. Like tossing a coin, the gadget comes up with a random "heads" or "tails" type decision, which means that on average it will be wrong 50 percent of the time. However, this is reckoned to be about the same level of accuracy attained by a typical human, but it avoids the stress of having to make the decision yourself!

This decision-maker has a seesaw action (Figure 4.32), with one end representing "heads" and the other end representing "tails". When the program is run the unit goes up and down a random number of times, and the end that is higher once the unit has finished provides you with the decision. It operates by using two touch sensors to determine when the moving platform has reached the limits of its movement, and the switches are used to reverse the direction of the motor. This gives the up and down movements as the motor driving the platform goes first one way, and then the other. Software is used to give a random number of oscillations so that the required random result is obtained each time the unit is activated.

Step 1 (Figures 4.33 - 4.36)

Start construction by building the rostrum. This is based on two identical (not "mirror image") pieces, and one of these is shown in Figure 4.33. To make

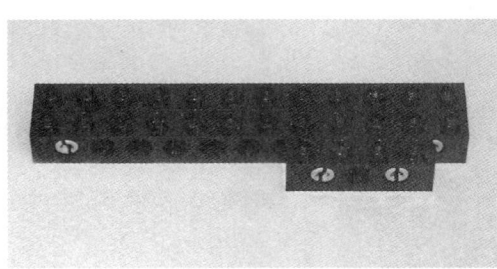

each one start by joining two 12 x 1 beams side by side using a peg at each end. Then fit two pegs into the end holes of a 4 x 1 beam, and fit it onto the 12 x 1 beams. Once you have two of these assemblies, a 4 x 1 beam and various plates are added to complete the base section. The parts required are shown in

Fig.4.33 One chassis side section

Figure 4.34. The top and bottom sides of the finished base section are shown in Figures 4.35 and 4.36.

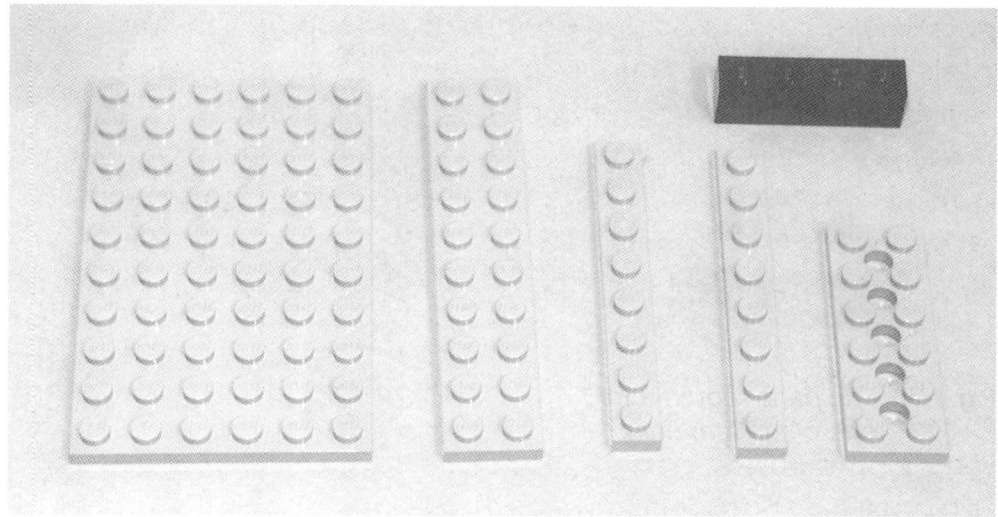

Fig.4.34 The parts needed to complete the base section

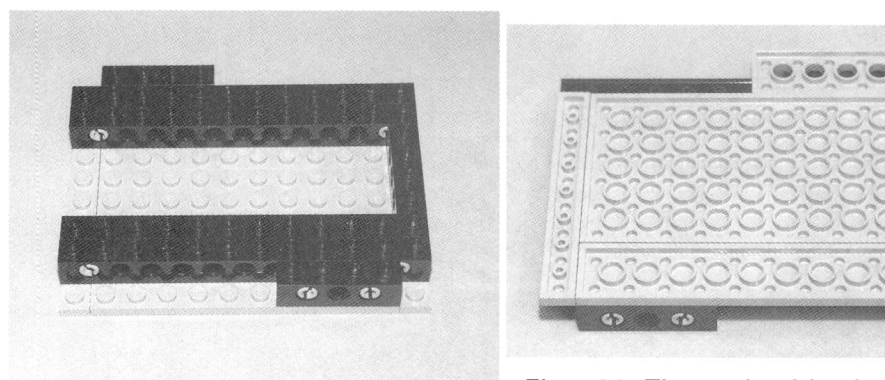

Fig.4.35 The top of the base section

Fig.4.36 The underside view of the base section

Step 2 (Figures 4.37 - 4.39)

The base must be fitted with two arms that limit the travel of the moving platform, and provide a reliable means of operating the touch sensors. The spring loading in the switches gives a simple but effective form of shock absorption, and helps to prevent the unit from self-destructing. Figure 4.37 shows the parts required for one arm assembly, and Figure 4.38 shows a

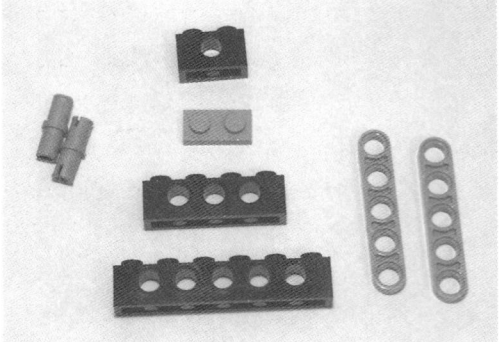

Fig.4.37 *The parts required for one of the arm units*

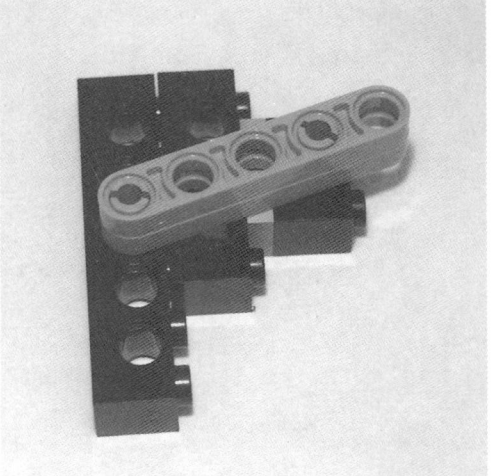

Fig.4.38 *A completed arm assembly*

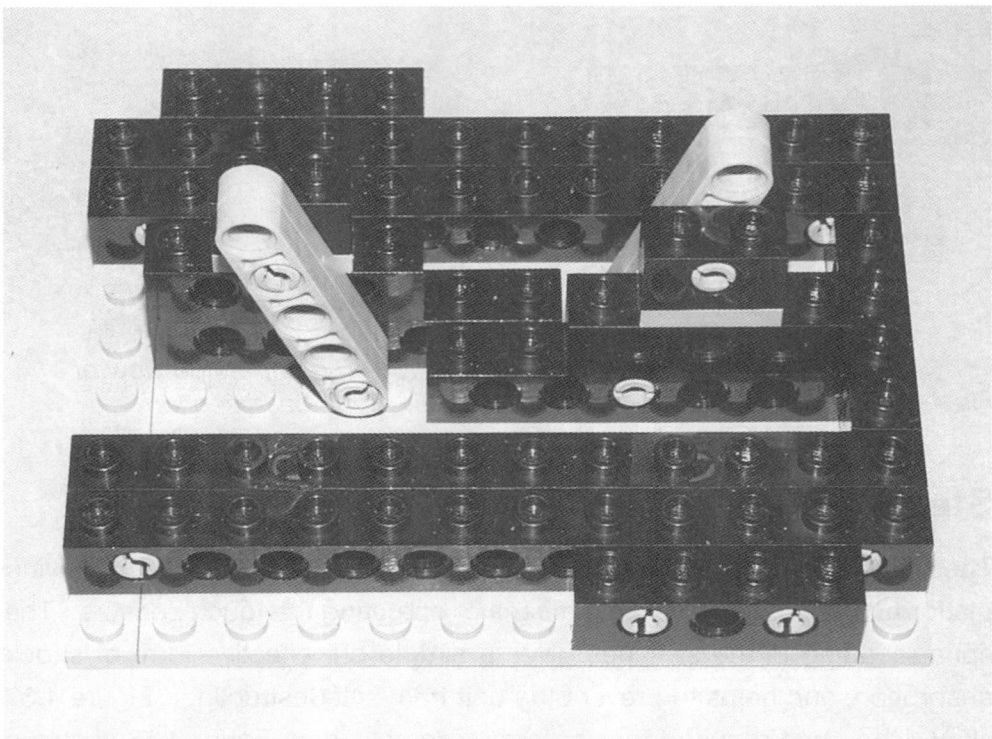

Fig.4.39 *The base section fitted with both of the arm assemblies*

completed assembly. Start by fitting a 4 x 1 beam at one end of and on top of a 6 x 1 beam. Then fit a 2 x 1 plate centrally on the 4 x 1 beam, and then place a 2 x 1 beam on top of the plate. The two 5 x 1 half thickness girder pieces are "riveted" together. The "rivets" go through the holes at one end of each girder, and the holes one in from the other end of the girders. Half of each "rivet" will be left protruding from the girder pieces, and these protrusions are used to fix the girder assembly to the beams, as per Figure 4.38. When the two arm assemblies have been completed they are fitted onto the base (Figure 4.39).

Step 3 (Figures 4.40 and 4.41)

The base section is now ready to take the two towers that support the moving platform. These are made from yellow girder pieces that have two 135-degree angles (Figure 4.40). The pieces required for one tower are shown in Figure 4.41. The two girder pieces are joined at the top using a grey "rivet" that has the standard fitting at one end and an axle-like fitting at the other. The other "rivet" is a black type that has a sort of axle style fitting at both ends. Fit them in the outer holes and not in the middle hole of the three, as the latter is used to provide the pivot points for the moving platform. The other two grey "rivets" are used to peg the bottom of the tower to the base. These are again the type that has a standard fitting

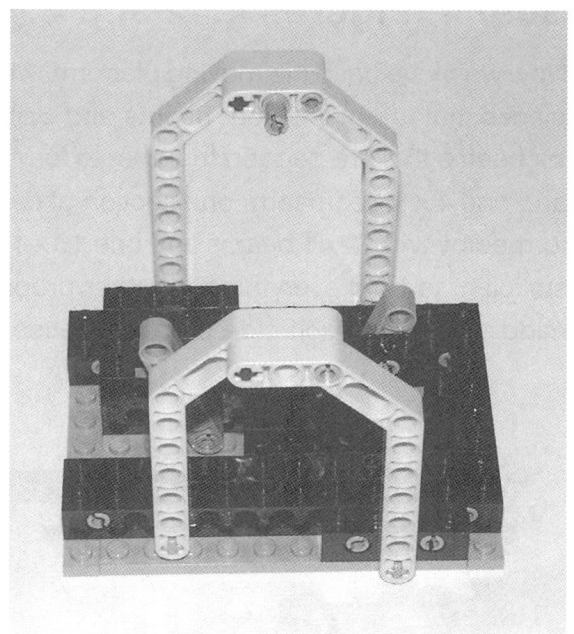

Fig.4.40 The towers in place on the base section

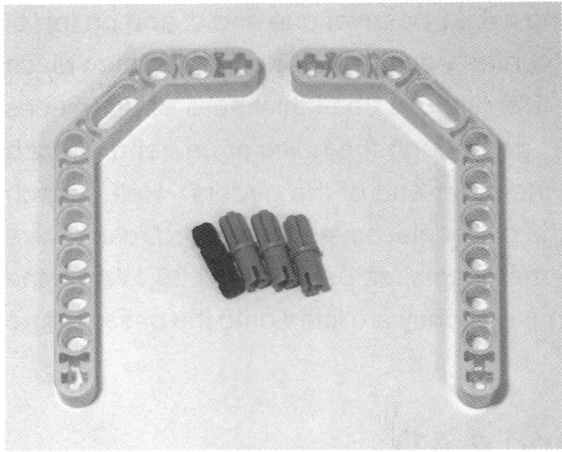

Fig.4.41 The parts needed for one tower

at one end and an axle-like fitting at the other. One tower is fitted with a standard "rivet" to act as one of the pivots for the platform. The hole in the other tower is left vacant, and an axle will eventually be fitted here. Do not worry if the towers seem rather flimsy. They are not masterpieces of engineering design, but they are still more than adequate to carry the platform.

Step 4 (Figures 4.42 and 4.43)

Now we move on to the moving platform, which is based on two identical side pieces that are made from beams and pegs. In effect, each side piece has six beams that are pegged together to form a single 36 x 2 beam. Two 16 x 1 and one 4 x 1 beam form one section of each piece, and the other section is formed by two 10 x 1 beams and one 16 x 1 type. About eight to ten pegs are sufficient to hold everything together properly. Do not use any pegs in the middle holes of each side piece, because these must be left free to act as

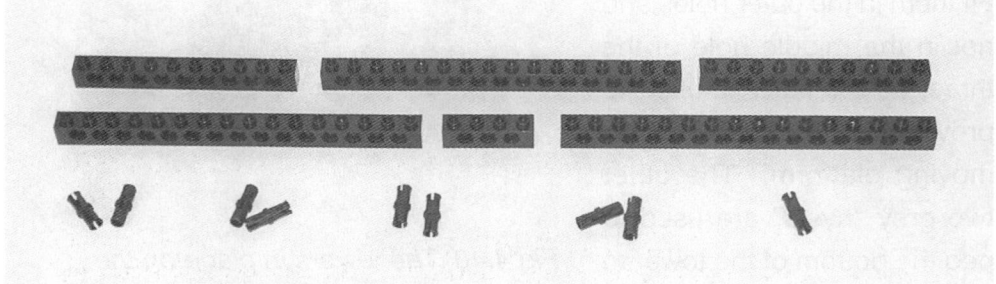

Fig.4.42 The parts required for one side of the platform

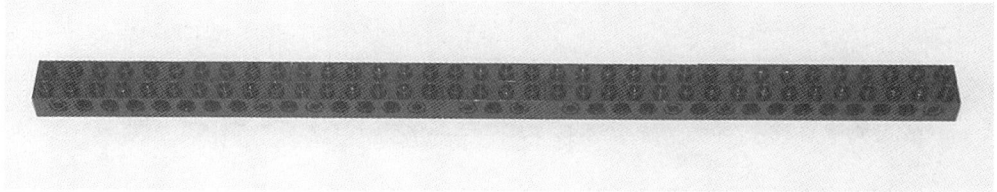

Fig.4.43 A completed side piece for the platform

pivot points. The parts for one side piece are shown in Figure 4.42, and the completed assembly appears in Figure 4.43.

Step 5 (Figures 4.44 - 4.46)

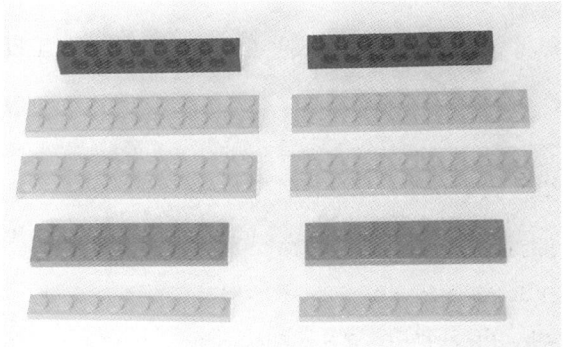

The two sides of the platform are now joined together using two 8 x 1 beams and various plates (Figure 4.44). Start by fitting the four 10 x 2 plates on the undersides of the side pieces, right at the ends, and protruding by a 2 x 1 area. These protrusions take the 8 x

Fig.4.44 The parts needed to join the two sides

1 beams, which then hold the two sides together. To further strengthen the assembly add the 8 x 1 plates on the underside (Figure 4.45), and the 8 x 2 plates on the top side (Figure 4.46).

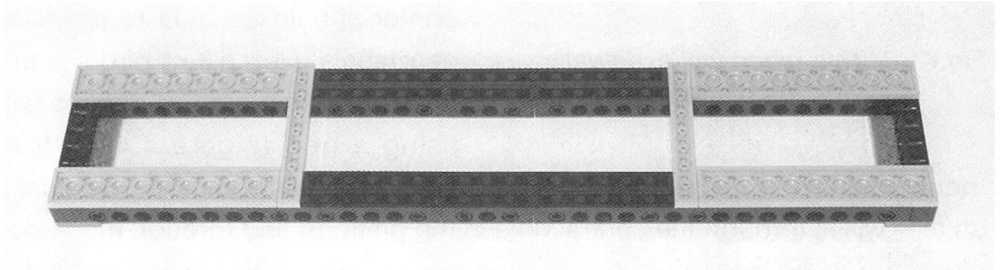

Fig.4.45 The plates on the underside of the platform

Fig.4.46 The top side of the moving platform

Fig.4.47 One of the legends

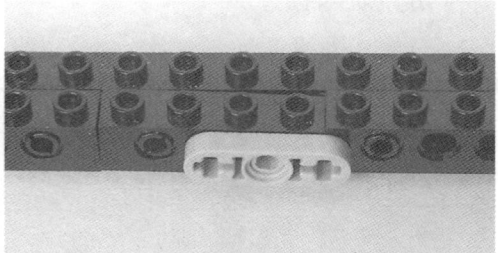

*Fig.4.48 This piece will eventually
lock the platform to an axle*

Step 6 (Figures 4.47 and 4.48)

One or two odds and ends are added to complete the platform. A 6 x 2 plate and an 8 x 1 beam are added at each end and the beams are fitted with the "heads" and "tails" labels (Figure 4.47). You can obviously use any wording you like on the labels, and I settled for "NO WAY!!!!!" and "GO FOR IT" on the prototype. I used an electronic labeller to make the labels, but you can obviously improvise here if necessary. Any PC equipped with a printer should be able to produce neat labels. Next a 3 x 1 girder piece is fixed on one side of the chassis using a peg (Figure 4.48). It is mounted one hole off-centre so that one of the end holes is central. Eventually an axle will fit through the central hole in the platform and through the girder piece, which will lock the platform to the axle. This enables the platform to be driven via the axle. The peg used should be one that has a short fitting at one end, and it is this end that fits into the girder piece. An ordinary peg leaves the girder piece quite loose. Finally, fit a 4 x 1 beam on one of the 8 x 1 plates to strengthen the platform slightly.

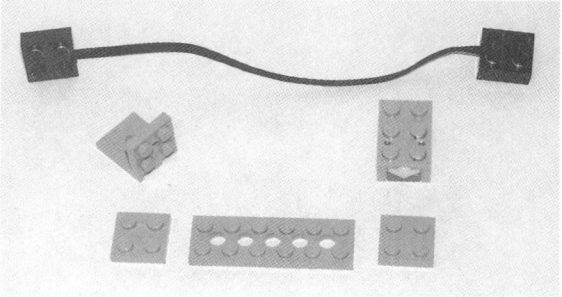

Fig.4.49   The parts for one sensor assembly

Step 7 (Figures 4.49 - 4.51)

Now the two sensor assemblies are constructed and fitted onto the platform. The parts required for one assembly are shown in Figure 4.49, and a completed assembly appears in Figure 4.50. Place the 2 x 2 plates

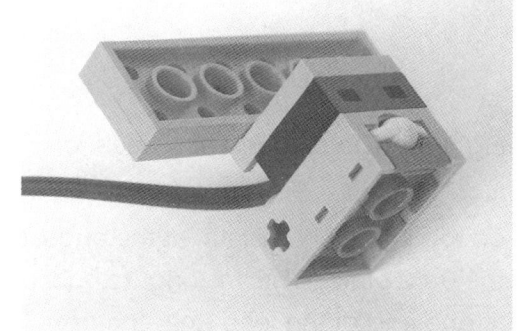

Fig.4.50   A completed sensor unit

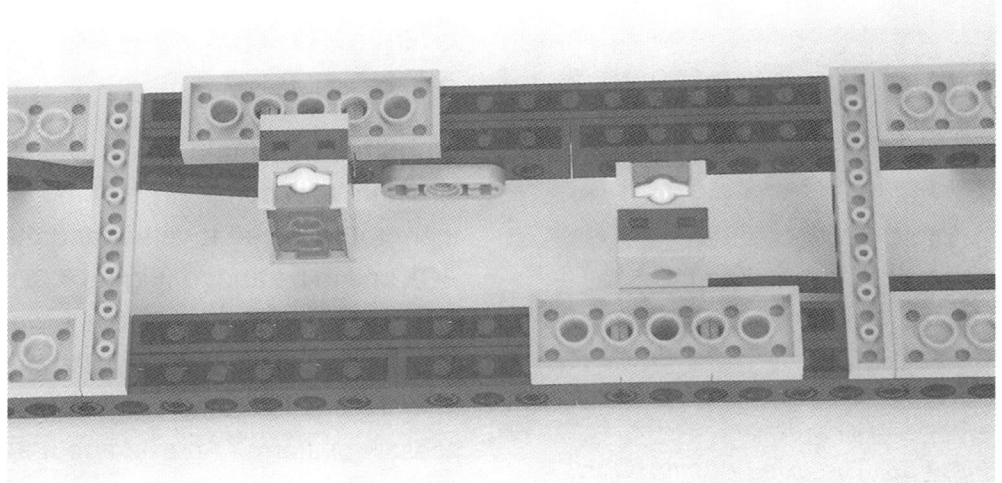

Fig.4.51   The two sensor assemblies fitted to the underside of the platform

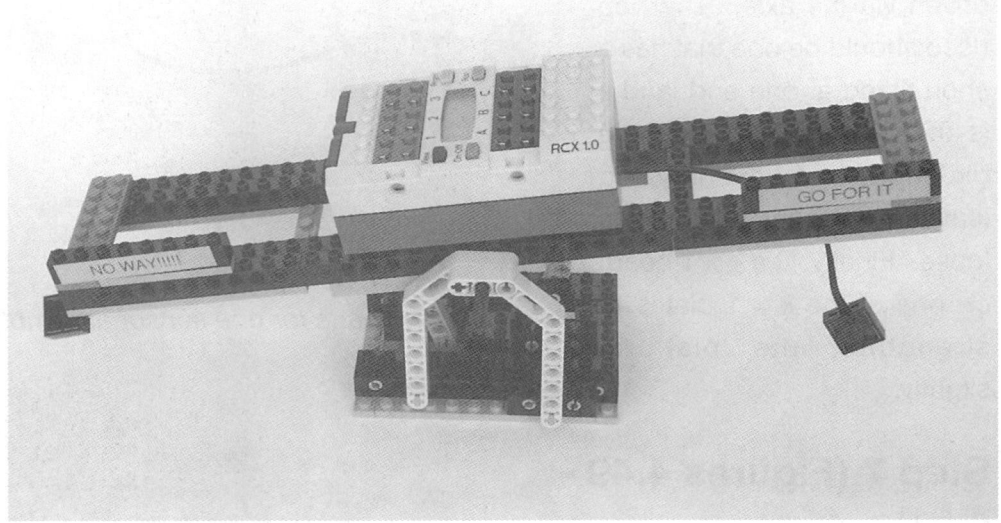

Fig.4.52 The platform and base joined, and the RCX unit fitted

on top of the 6 x 2 plate, at the ends, leaving a 2 x 2 area free in the middle of the 6 x 2 plate. The 4 x 2 right angle plate is fitted onto this central area. Next fit one end of the lead onto the front 2 x 2 area of the sensor, and then fit the connector onto the right angle plate. Repeat this process to make the other sensor assembly, and then fit them both onto the platform (Figure 4.51).

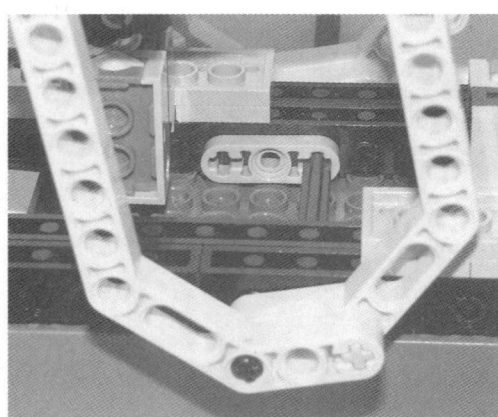

Fig.4.53 The axle goes through the small girder piece

Step 8 (Figures 4.52 and 4.53)

With the two main parts of the structure now complete it is time for them to be joined together and the RCX unit to be added (Figure 4.52). There is a peg already in one tower on the base section, and this fits into one of the holes at the middle of the seesaw platform. An axle about 94 millimetres long provides the pivot on the other side. This is carefully threaded through the relevant holes in the platform and the other tower. It must be pushed as far as it will go, so that it actually goes a little way into the tower that carries the first pivot. However, be careful not to push it so far that it displaces the peg that provides the first pivot. Also be careful not to displace the small grey piece fitted on the platform, that locks the axle to the platform (Figure 4.53). This will have to be held in place while the axle is pushed into position. With the platform fitted, the RCX unit is added to the platform.

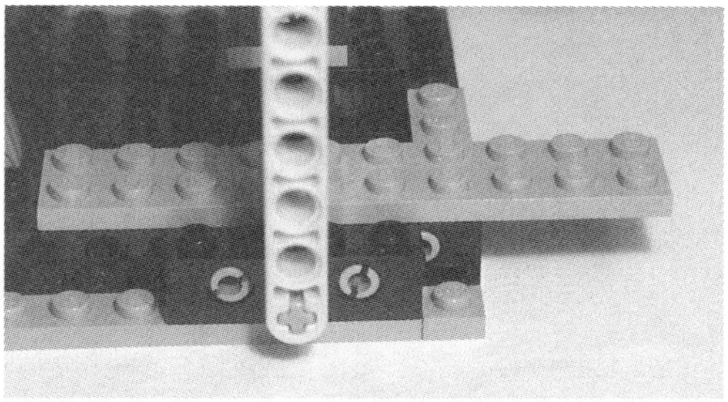

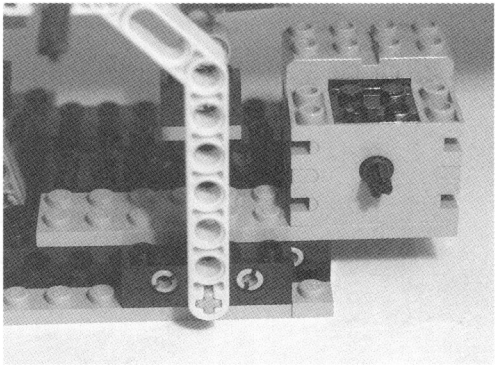

Fig.4.54 *The simple mounting pad for the motor*

Fig.4.55 *The motor in position on the mounting pad*

Step 9 (Figures 4.54 and 4.55)

Now the motor is added, and this requires a mounted pad to be fitted on the base section of the unit. The pad just consists of a 2 x 1 plate and a 10 x 2 type (Figure 4.54), to which the motor is then added (Figure 4.55).

Step 10 (Figures 4.56 and 4.57)

Next the pulleys and driving bands are installed, and the parts needed are shown in Figure 4.56. A large step-down ratio is needed because the unit will otherwise operate so fast that it will probably "self-destruct in five seconds". Therefore a two-stage reduction drive is used, with a pulley of the smallest size driving the largest type in both cases. This gives a total reduction ratio of about 30 to 1. One of the small pulley wheels is fitted onto the spindle of the motor. An axle about 31 millimetres long is pushed through the fifth hole up

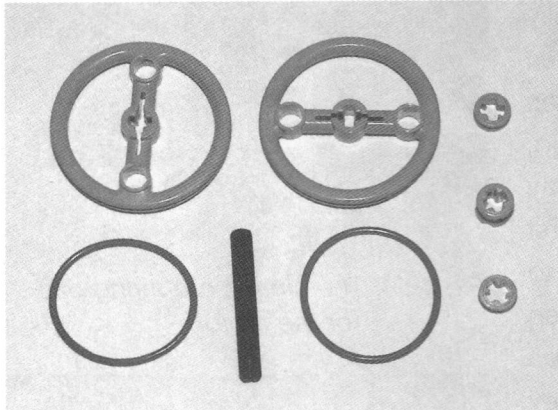

Fig.4.56 *The parts needed for the drive assembly*

Fig.4.57 *The completed drive system*

in the girder piece near the motor, and a large pulley wheel is fitted onto it. It is a good idea to fit the driving band onto the pulley before it is fitted on the axle, since it is awkward to fit the band afterwards. Both driving bands are the blue type about 25 millimetres in diameter. The small pulley is then fitted at the other end of the axle, and the driving band is stretched over the large pulley and the small one on the motor. If necessary, move the small pulley wheel along the motor's spindle to get the alignment of the two wheels just right. The second large pulley wheel is then fitted onto the axle that is already in place at the top of the towers, and a fixing "nut" is added to help keep it in place. The second driving band is then fitted, giving the finished assembly of Figure 4.57.

Step 11 (Figures 4.58 and 4.59)

To complete the unit the leads from the touch sensors are connected to inputs 1 and 3 of the RCX unit, and the motor is connected to output A (Figure 4.58). The sensors are already fitted with leads, but an additional lead is

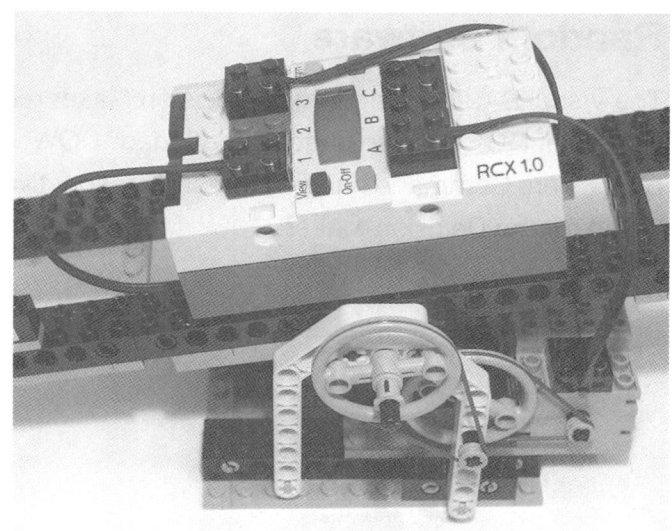

Fig.4.58 The leads connected to the RCX unit

needed to connect the motor to the RCX unit. This gives the finished unit of Figure 4.59

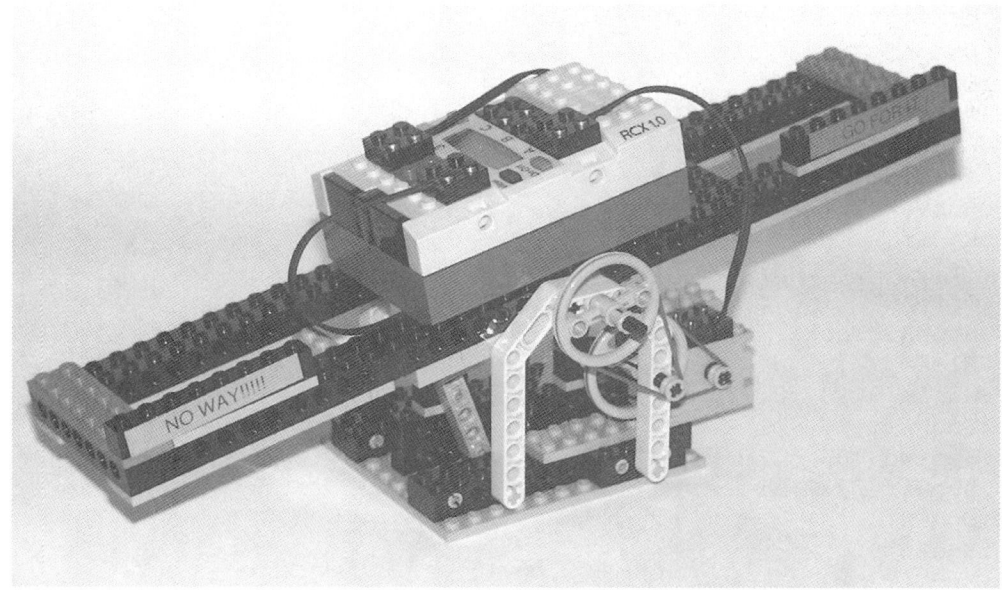

Fig.4.59 The completed executive toy, ready for testing

Random software

The Visual BASIC listing for the Decision Maker requires one form, complete with Spirit.OCX and two buttons captioned "DOWNLOAD" and "EXIT". These buttons are Command1 and Command2 respectively. This is the full software listing for the Decision Maker:

```
Private Sub Command1_Click()
With Spirit1
.InitComm
.SelectPrgm 1
.BeginOfTask 0
.SetSensorType 0, 1
.SetSensorMode 0, 1, 0
.SetSensorType 2, 1
.SetSensorMode 2, 1, 0
.SetVar 2, 4, 14
.SumVar 2, 2, 6
.On "0"
.SetRwd "0"
.Loop 2, 0
.If 0, 2, 2, 2, 0
.Off "0"
.StopAllTasks
.Else
.While 9, 2, 2, 2, 0
.EndWhile
.SetFwd "0"
.SubVar 2, 2, 1
.EndIf
.If 0, 2, 2, 2, 0
.Off "0"
.StopAllTasks
.Else
.While 9, 0, 2, 2, 0
.EndWhile
.SetRwd "0"
.SubVar 2, 2, 1
.EndIf
.EndLoop
.EndOfTask
End With
End Sub
```

```
Private Sub Command2_Click()
Spirit1.CloseComm
End
End Sub
```

The program is organised as a single task, and it starts by setting up inputs 1 and 3 to operate with the touch switches in Boolean mode. Consequently, the value returned from an input is 0 under standby conditions, or 1 when a sensor switch is activated. A random value is needed to control the number of oscillations that the device performs, and a SetVar instruction is used to place a random number from 1 to 14 in variable 2. We need the unit to perform a minimum number of oscillations that is greater than 1, so 6 is then added to variable 2 using a SumVar instruction. This gives a random number from 7 to 20, but this is the number of half cycles that the platform will go through. The unit therefore performs between 3.5 and 10 up/down cycles. Next the motor is switched on and set in reverse, and this sends the platform tilting towards the sensor on input 3.

The program then goes into an infinite loop, followed by an If…Then…Else structure, which tests the value in variable 2 and acts on the result. The value in this variable is decremented each time the platform completes a half cycle, and when the value reaches zero the If instruction causes the following two commands to be performed. These switch off the motor and then end all tasks. Note that although the program contains an indefinite loop, the loop and the entire program will actually be brought to a halt when the platform has gone through the appropriate number of half cycles. On the first run through the value in variable 2 will be higher than zero, and the instructions after the Else command will be performed. The first of these is a Do…While loop that actually does nothing while the value returned from input 2 is zero. The purpose of this loop is to wait for the value from the sensor to change. Before long the platform activates the touch switch on this input, the returned value changes to 1, and the program moves on. The motor is then switched to go forwards, and a SubVar instruction deducts one from the value stored in variable 2. The If…Then…Else structure is then terminated.

Now the program goes into another If…Then…Else structure that is virtually identical to the first one, but the Do…While loop monitors the touch switch connected to input 1. Before too long this sensor is activated, the motor is returned to reverse operation, and one is again deducted from the value stored in variable 2. The program then goes back to the beginning of the loop. This process repeats until the value in variable 2 is reduced to zero, and the program then terminates when the next If instruction is reached. To make another decision simply press the Run button on the RCX unit.

Finally

True randomness is quite hard to achieve using computer circuits, but this unit seems to achieve a passable simulation of tossing a coin. It may occasionally come up with the same answer several times in succession, but this can also happen when tossing a coin. You might like to try altering the program to produce the computerised equivalent of a "double-headed" coin.

Direct control

Basic control

So far we have been concerned with robots that run an onboard program which enables them to function without any human intervention. This is the most interesting and challenging way of controlling a robot, but there is a lot of play value in using some form of remote control. The infrared link between the RCX unit and the PC gives more than ample range to make remote control from the PC a practical proposition. However, remember to set the switch on the infrared transmitter for long distance operation (set the slider towards the elongated triangle symbol).

As an initial experiment with remote control via a PC, start Visual BASIC, and then slightly enlarge the Project window and the form within it. Alter the name of the form and its caption to something suitable, such as RemCont. Add the Spirit.OCX control and then put three control buttons across the middle section of the form. Change the captions of these buttons to FORWARD, REVERSE, and STOP. Add a fourth button below these and change its caption to EXIT. The screen should then look something like Figure 5.1.

Double-click on the form to bring the Code window to the "front", and then add this line of code:

```
Spirit1.InitComm
```

This simply initialises the serial port for operation with the infrared transmitter when the program is run. Next add these two lines for the FORWARD button:

```
Spirit1.On "02"
Spirit1.SetFwd "02"
```

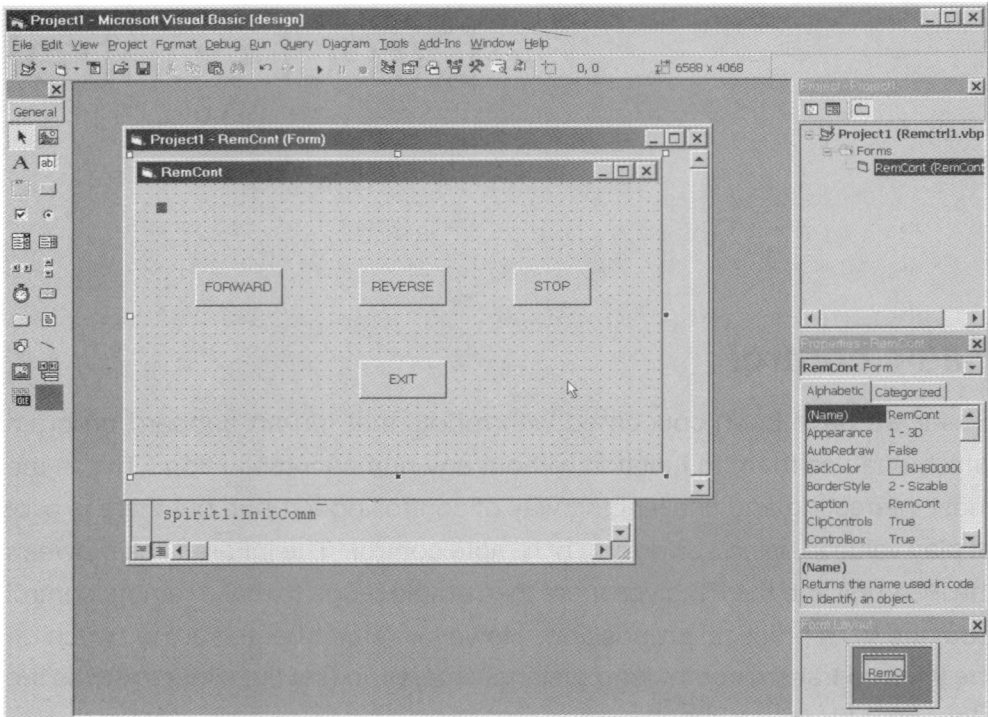

Fig.5.1 The form set up for the remote control program

These two commands are transmitted to the RCX unit when the FORWARD button is operated, and they turn on outputs 1 and 3 and set both of these outputs to forwards operation. Add these two lines of code for the REVERSE button:

```
Spirit1.On "02"
Spirit1.SetRwd "02"
```

These two commands are transmitted when the REVERSE button is operated, and they set outputs 1 and 3 on and for reverse operation. Next add this line for the STOP button:

```
Spirit1.Off "02"
```

This command simply switches off output 1 and 3 when the STOP button is operated. Finally, add these two lines for the EXIT button:

```
Spirit1.CloseComm
End
```

When the EXIT button is operated the first of these lines closes communications with the infrared transmitter, and then the second line closes the program. Assuming the FORWARD, REVERSE, STOP, and EXIT buttons are Command1 to Command4 respectively, the completed program listing should look like this:

```
Private Sub Command1_Click()
Spirit1.On "02"
Spirit1.SetFwd "02"
End Sub

Private Sub Command2_Click()
Spirit1.On "02"
Spirit1.SetRwd "02"
End Sub

Private Sub Command3_Click()
Spirit1.Off "02"
End Sub

Private Sub Command4_Click()
Spirit1.CloseComm
End
End Sub

Private Sub Form_Load()
Spirit1.InitComm
End Sub
```

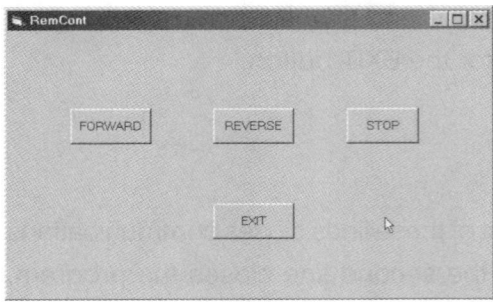

Fig.5.2 The remote control program in operation

This program provides basic control of any rover style robot that uses drive motors connected to output 1 and 3 of the RCX unit, including those featured in the two previous chapters. A window having the four control buttons should appear when the program is run (Figure 5.2), and the robot should respond in the appropriate fashion to the three control buttons. The EXIT button has not been set up to switch off the motors before closing the program, but you can simply switch off the RCX unit if you inadvertently exit the program before the motors are stopped. Alternatively, you can start the program again, turn off the motors, and then exit the program.

The infrared link is quite effective, and signals reflected off walls, objects in the room, etc., are usually sufficient to provide proper communications between the PC and the RCX unit. Consequently, communication is not dependent on the transmitter being aimed at the front of the RCX unit. On the other hand, if you give the transmitter a "clear shot" at the RCX unit the chances of everything working well are maximised. It the system is used out of doors or in a very large room there will be a lack of reflected infrared "light" to assist communications, and results may be relatively poor.

Steering

This program only provides very limited control, but the program is easily modified to add more features. Click on the EXIT button if the program is still running, and then add two more control buttons above the FORWARD and STOP button. Change the captions of the new buttons to LEFT and RIGHT. Double-click on the LEFT button and assign this code to it in the Code window:

```
Spirit1.SetFwd "2"
Spirit1.SetRwd "0"
```

Operating this button sets the motor on output 1 into reverse and the motor on output 3 to go forwards, which turns the robot to the left. Next add this code for the RIGHT command button:

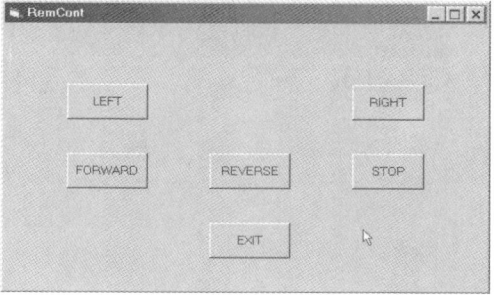

```
Spirit1.SetFwd "0"
Spirit1.SetRwd "2"
```

Fig.5.3 The improved remote control program

The RIGHT button has the opposite effect to the LEFT one, placing the motor on output 3 into reverse and setting the one on output 1 to go forwards. This turns the robot to the right. When the program is run you should get a window something like Figure 5.3, and with the aid of the new buttons it is then possible to steer the robot.

Speed control

These buttons give good control over the robot, but precise control can be difficult if the robot is a fast type. It can be particularly difficult if the robot is one that can turn very rapidly. A speed control is a decided asset, and is easily added. There is more than one way of handling this, but the obvious one is to use a slider control, which is actually a form of Scrollbar in Visual BASIC terminology. In the toolbox at the left-hand side of the screen you will find two forms of scrollbar available, which are the horizontal and vertical varieties. You will probably be able to pick them out from their icons without too much difficulty, but if you position the pointer over an icon its name will pop up. If you are at all unsure, try placing the pointer over the icons, one by one, until you find the scrollbars. It does not matter which one we use in this case, but I would suggest using the horizontal type as it will fit nicely in the space between the LEFT and RIGHT command buttons.

Having left-clicked on the horizontal scrollbar icon you then drag a rectangle on the form using the mouse, and the scrollbar will then appear on the form. If necessary it can be dragged to a new position and the handles can be used

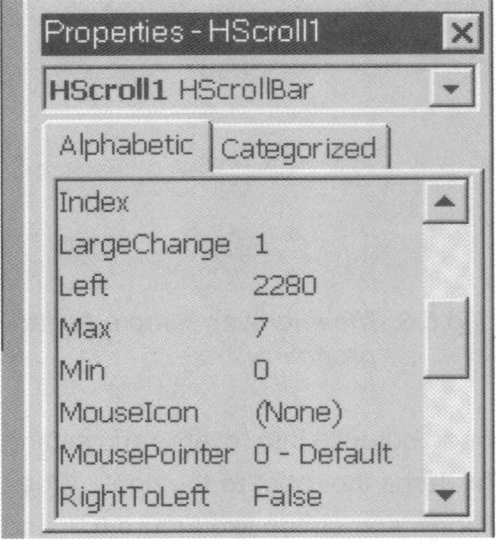

*Fig.5.4 Changing the scrollbar's
Max setting*

to change its size. I would not bother making it very large since there are only eight power levels, and fine control of the robot's speed is not a problem. A full range of values for the scrollbar will appear in the Properties window, but most of these can simply be left at the default values. One exception is the range of values provided by the control. By default these range from a minimum of zero with the control set fully to the left, to a maximum of 32767 with it set as far as possible to the right. For our purposes the minimum value of zero is fine, but the maximum value must be changed to 7 (Figure 5.4).

The Small Change and Large Change values will probably change to 1, but check they are at this value. These are respectively the increments provided when dragging the slider control and moving the control by left clicking just to the left or right of the "control knob". The clicking method is normally set to give a larger change, but with a range of only eight values available there is no point in having either of these at anything other than 1.

Next double-click on the slider control to bring up the Code window, and then add this line of code for the control:

```
Spirit1.SetPower "02", 2, HScroll1.Value
```

Changing the setting of the control causes this command to be sent to the RCX unit. It sets the power of outputs 1 and 3, and the second parameter tells the command to use the constant value provided by the third parameter. Constant is not really the right term in this example, because the command actually uses the new value read from the control. This value is in the variable called HScroll1.Value. Changing the setting of the control therefore results in a SetPower command being transmitted to the RCX unit, with the new power value being read from the control.

The scrollbars do not have a caption property, but the label object can be selected from toolbox, and then a label can be added above the control. Its caption can be changed to SPEED using the Properties window.

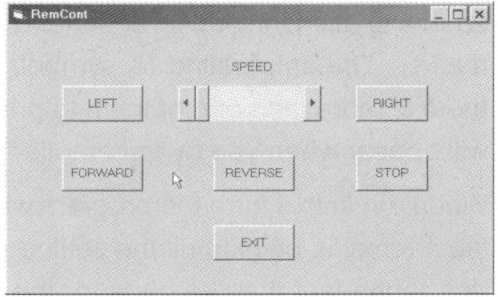

Fig.5.5 The speed control added to the remote control program

When the program is run you should get a window that looks something like Figure 5.5. The initial setting of the SPEED control is 0 by default, but you can set a different power level by changing its Value setting in the Properties window. The program now has a full range of controls, and provides excellent control over the robot. One possible improvement is to have the label indicate the actual power setting, and this is quite easy to achieve. Start by left clicking on the label to select it, and then delete the existing caption via the appropriate section of the Properties window. Next double-click on the scrollbar to bring up its section of the code window, and add this line of code after the existing line:

```
Label1.Caption = "SPEED IS " & HScrollBar1.Value
```

Although one might expect the code for the label to be placed in its section of the code window, it is when the control is altered that the label must be updated. Hence this code is applied to the scrollbar. The original line of code sends the appropriate message to the RCX unit each time that the control is adjusted, and then the new line updates the label. The text in the double quotes is printed on the label, followed by the value in the variable that contains the current control setting. We require a space between the text and the value, so this space is included at the end of the text within the double quotation marks. The ampersand (& symbol) has to be used between items where there is more than one of them to print. If you leave it out an error message will appear when you try to compile the program.

When run in this form the program will almost work, but you will find that the label remains blank until the setting of the speed control is altered. This is due to the fact that the code for the control is only run when the control is altered, which is all we need to provide speed changes. In order to make the label appear as soon as the program is run it is merely necessary to add another line of code. This time it is applied to the form, so left-click on the form to bring it to the "front" and then double-click on the form to go to its section of the Code window. Then add this line after the existing line of code:

```
Label1.Caption = "SPEED IS " & HScrollBar1.Value
```

This is exactly the same as the one applied to the scrollbar previously, and it has the same effect. As soon as the program is run and the form is loaded, this program line prints the label on the screen, complete with the initial speed setting. Another improvement to the program is to add a line here to send an initial speed setting to the RCX unit. For example, if the scrollbar were set for an initial value of 4, this program line would send an initial power setting with a value of 4.

```
Spirit1.SetPower "02", 2, HScrollBar1.Value
```

If you run the program it should look something like Figure 5.6. No doubt further refinements could be added. Another label could be included and used to show the currently selected setting (FORWARD, STOP, etc.). This would require a new program line to be added for each of the buttons, with the obvious exception of the EXIT button. As another little programming exercise you could try adding a command button to make the RCX unit "beep" each time the new button is operated.

When using a PC to provide a remote control function, do not be surprised if things happen something less than instantly. The infrared link to the RCX unit is reasonably fast. However, there is a small but perceptible delay between an onscreen control being operated and the robot responding to the change.

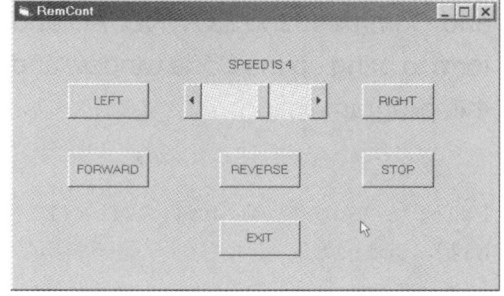

Syntax

If you look through the lines of code provided so far in this chapter you will notice that some of them have

Fig.5.6 The final version of the remote control program

the Spirit1. prefix while others do not. The important point to keep in mind is that this prefix is only needed when using functions that utilize the Spirit.OCX control. The prefix is not necessary and must not be used for program lines that used standard Visual BASIC commands.

Another point to note is that many of the commands provided by Spirit.OCX are not applicable when controlling the RCX unit in immediate mode. In particular, those commands that control the program flow are not usable in the immediate mode, because you are running a Visual BASIC program on the PC, and there is no program running in the RCX unit, other than its built-in firmware of course. Program flow, where appropriate, is therefore controlled by the Visual BASIC program and not by the RCX unit.

Mixed operation

The RCX unit is not limited to operating in the immediate mode or having deferred operation, and it is possible to combine the two. Mixing the two methods inevitably gives more scope for things to go wrong, and for this type of thing to work well it is necessary for the program running in the RCX unit to be written with partial manual control in mind. If you simply send commands to a robot running a normal control program things may well go awry. As a simple demonstration of mixed mode operation, start by placing Spirit.OCX and two command buttons on a form. Change the captions for Command1 and Command2 to DOWNLOAD and EXIT respectively. Double-click on the form to bring up the Code window and then add the necessary lines to produce this program listing:

```
Private Sub Command1_Click()
With Spirit1
.InitComm
.SelectPrgm 2
.BeginOfTask 0
.On "02"
.SetFwd "02"
.SetSensorType 0, 1
.SetSensorMode 0, 1, 0
.SetSensorType 2, 1
.SetSensorMode 2, 1, 0
.StartTask 1
.StartTask 2
.EndOfTask

.BeginOfTask 1
.Loop 2, 0
.If 9, 0, 2, 2, 1
.SetRwd "02"
.Wait 2, 100
.SetFwd "0"
.EndIf
.EndLoop
.EndOfTask

.BeginOfTask 2
.Loop 2, 0
.If 9, 2, 2, 2, 1
.SetRwd "02"
.Wait 2, 100
.SetFwd "2"
```

```
.EndIf
.EndLoop
.EndOfTask
End With
End Sub

Private Sub Command2_Click()
Spirit1.CloseComm
End
End Sub


Private Sub Form_Load()

End Sub
```

I will not describe this program in detail, because it is just a simplified version of the Ditherbot program from chapter 4. The difference here is that the random element has been removed. After detecting an obstruction the robot will reverse for one second, as it did using the original software. The robot then starts to turn, but rather than turning for a random length of time it just keeps turning indefinitely. In order to break it out of its pirouette a suitable command in immediate mode must be sent to the RCX unit. A very simple control program could be written to do this, but the remote controller program featured previously in this chapter will do the job perfectly well.

If you try out this set up, when the robot gets into a spin, wait until it is pointing in the desired direction and then operate the FORWARD button on the screen. This should break the robot out of the spin and set it going forwards again. In fact all the onscreen controls should work, and the only difference to normal remote controlled operation is that the robot will go into its automatic reversing and spinning routine if it collides with something.

Reading the RCX unit

The infrared link enables the RCX unit to send data back to the PC, which makes it possible to read the RCX unit's inputs, plus some other information such as the battery voltage. Commands such as If and SetVar are only usable in programs that are downloaded to the RCX unit, and they are not applicable to operation in immediate mode. As a simple introduction to reading the RCX unit, try this simple program for reading the battery voltage:

```
Private Sub Command1_Click()
Label1.Caption = Spirit1.PBBattery
End Sub

Private Sub Command2_Click()
Spirit1.CloseComm
End
End Sub

Private Sub Form_Load()
Spirit1.InitComm
End Sub
```

A form having two command buttons is required, and these should be labelled READ (Command1) and EXIT (Command2). A label is also required, and Spirit.OCX should obviously be included on the form. The line of code attributed to the form simply opens communications with the infrared transmitter when the program starts up. Operating the READ button uses the PBBattery command to read the battery voltage from the RCX unit, and it is used as the caption for Label1. Operating the EXIT button closes communications with the infrared transmitter and ends the program.

If you try the program you should obtain something like Figure 5.7. Operate the READ button in order to get a new battery reading. The battery reading is in millivolts, and must be divided by 1000 in order to obtain a reading in volts. The reading is an average over the previous 30 seconds, so the system will not be "fooled" into giving a very low reading due to the battery happening to be heavily loaded at the instant a measurement was made. With fresh batteries the reading should be around 9000, even with a couple of motors running. The reading of 7676 shown in Figure 5.7 was obtained with no motors being driven, and indicates that the batteries are nearing the end of their operating lives. Incidentally, you soon know when the battery voltage is no longer adequate, because the RCX unit immediately shuts down!

Fig.5.7 The battery monitor

It is possible to make some minor but useful improvements to the way data is displayed. In Figure 5.7 the size of the text on the buttons and label has been enlarged. In order to do this, select the item containing the text and the go to the Font entry in the Properties window. Left-click on the word Font to select it, and then left-click on the button that will appear just to its right. This will bring up a

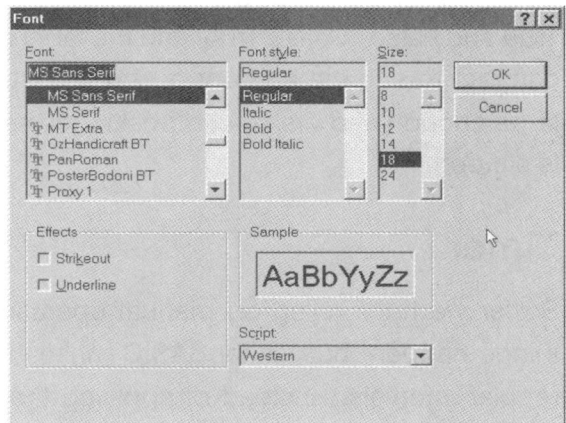

Fig.5.8 Editing the font settings

window like Figure 5.8 where you can select various font sizes and styles, or even change to a new font. In order to have the display read in volts rather than millivolts, replace the original program line attributed to the READ button with this one:

```
Label1.Caption = (Spirit1.PBBattery / 1000) & " VOLTS"
```

As before, we are using the PBBattery command to read the battery potential, and we are using the reading as the caption for Label1. However, before

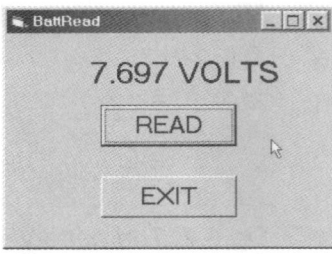

Fig.5.9 The modified battery monitor

printing the reading on the screen it is divided by one thousand so that it is in volts rather than millivolts. Also, the word VOLTS has been added after the reading. Note that there must be a space ahead of the word VOLTS so that it is spaced slightly from the figures in the reading. The displayed readings should then look something like the one in Figure 5.9. If you find that some of the text is missing, exit the program, go to the form, and make the label wider. Alternatively, select the label, go to the Properties window, select Autosize, and then left-click on the button that appears to the right of this entry. This will produce a pop-down menu having just two options, which are True and False. Select True, and the label will then automatically expand and contract to suit the amount of text it contains. Provided the form is made large enough and the label is positioned sensibly, this will ensure that no text is clipped.

Timer

So far we have relied on manual operation of command buttons to make things happen, but Visual BASIC makes it easy to make things happen at regular intervals. Instead of applying the code to a command button it is assigned to a timer component. To demonstrate the user of a timer we will produce an automatic version of the battery voltage monitor. By using a timer component this version automatically updates the display once per second. Start by selecting the READ button and then pressing the Del key to delete it from the form. Next select the timer component from the toolbox down the left-hand side of the screen. You can not really miss this one as it looks like a stopwatch. Drag a rectangle onto the form, and the timer icon will then

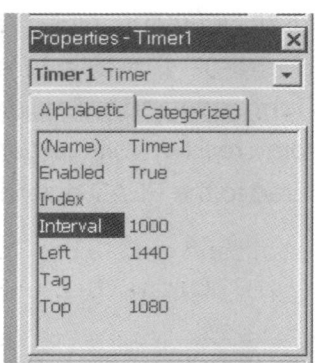

Fig.5.10 Changing the timer interval

appear on the form. The positioning and size of the timer icon is unimportant, because it will not appear when the program is run. Double-click on the timer to bring up its section of the Code window, and then add this line of code:

```
Label1.Caption = (Spirit1.PBBattery / 1000) & " VOLTS"
```

This is exactly the same line that we used previously for the READ button, and it has the same effect. Before running the program you must set the required interval for the timer. Left-click on the timer to select it, and then go to the Properties window at the right-hand side of the screen. The interval setting is in milliseconds, and is 0 by default. A value of 0 disables the timer incidentally. We require a one-second delay so this setting must be altered to 1000 (Figure 5.10). When the program is run you should get something like Figure 5.11, with no timer icon in the window, but the display being automatically updated once per second.

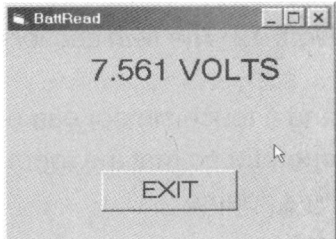

Fig.5.11 The new battery monitor program

Feedback

It would presumably be possible to have an autonomous robot that was controlled via a program running on the PC rather than in the RCX unit, but there is no obvious reason for doing things this way. Just the opposite in fact, because having the control program running in the RCX unit avoids any problems with delays in the swapping of data with the PC. It also avoids problems due to breaks in the link with the PC. Last and by no means least, once programmed the robot can be used practically anywhere, and not just where there happens to be a suitable PC handy.

A more interesting line of experiment is to have the robot send data to the PC so that the robot can be driven "blind". As most readers will no doubt be aware, industrial robots are often used in inaccessible or dangerous places where human workers can not venture. Robots of this type are normally operated by remote control, and the robot has to send back information to a

Fig.5.12 The twin sensor of Feedbackbot

human controller. He or she can then use this information to accurately perform the necessary tasks. The information from the robot is usually provided by video cameras, possibly backed up by other sensors. For the time being this sort of thing is well beyond the Lego MindStorms kits. However, a light sensor and a touch sensor can be used to send back some basic information to the operator so that the robot can perform a simple task, such as homing in on a light source.

The simple rover style robot described here (Feedbackbot) is equipped with a light sensor and a touch sensor at the front (Figure 5.12), and it is suitable

Fig.5.13 A rear view of Feedbackbot showing the motor assembly

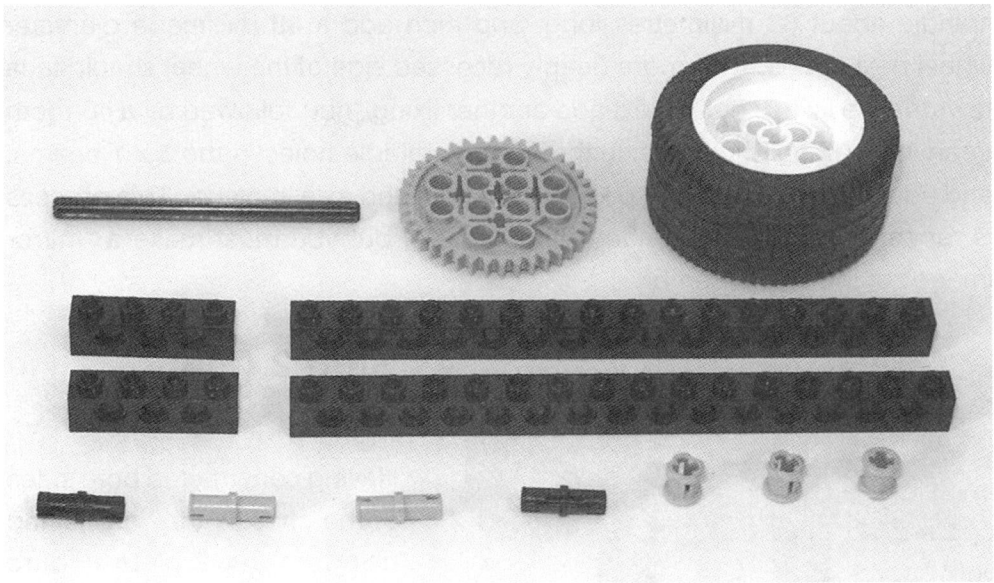

Fig.5.14 The components needed to build one side of the chassis

for experiments with "blind" operation of a robot. Figure 5.13 shows the rear view of the robot.

Step 1 (Figures 5.14 and 5.15)

The first step is to construct the two halves of the chassis, complete with the rear wheels. Figure 5.14 shows the parts needed for one side of the chassis, and Figure 5.15 shows a completed assembly. Two pegs are used to join the

Fig.5.15 A completed side assembly

two 16 x 1 beams side by side. Fit the pegs into the end holes in the beams. Repeat this process with the 3 x 1 beams, and then fit them under the 16 x 1 beams, right at one end of these beams. Place a fixing "nut" on the end of a spindle about 63 millimetres long, and then add a 49 millimetre diameter wheel next to this. The more deeply recessed side of the wheel should face towards the fixing "nut". Next add another fixing "nut" followed by a 40-tooth gearwheel. Finally, fit the axle through the middle holes in the 3 x 1 beams, and add a third fixing "nut" to keep the wheel and axle in place. This process is repeated for the other side of the chassis, but you must make a "mirror image" of the first assembly.

Fig.5.16 *The two sides joined together*

Step 2 (Figure 5.16)

Having constructed both sides of the chassis they are joined using an 8 x 2 plate (Figure 5.16). This does not hold the two halves together very well, and more plates can be used if desired. However, as more items are added to the chassis the bond between two halves becomes much stronger.

Fig.5.17 *The full front wheel assembly, complete with both sensors*

Step 3 (Figures 5.17 - 5.19)

The chassis is now ready for the front wheels and the sensors to be added. These are built as a single assembly (Figure 5.17), but we will take assembly of this unit in two steps. The front wheel assembly (Figure 5.18) is built first, and then the sensors are added. The parts needed for one wheel assembly are shown in Figure 5.19, together with an assembled unit and the plate that joins the two completed wheel units together. Start by fitting a fixing "nut" at the end of an axle about 47 millimetres long. Then add a wheel of about 42 millimetres in diameter (including the tyre), followed by another fixing "nut", pressing all three close together. Next fit two 2 x 1 beams onto a 2 x 2 plate. The beams must both be the type that has a hole for an axle, and they must be orientated so that the two holes line up. This assembly is added onto the axle, and a third fixing "nut" is

Fig.5.18 The front wheel assembly is built first. The sensors are added later

Fig.5.19 The parts required for the front wheel assembly

added to hold it in place. Repeat this process to produce the other wheel assembly, and then add them at opposite ends of an 8 x 2 plate.

Step 4 (Figures 5.20 - 5.22)

The sensors and other parts needed to complete the front wheel and sensor assembly are shown in Figure 5.20. Add the light sensor onto the 8 x 2 plate with its rear 2 x 2 area in the middle of the plate. There

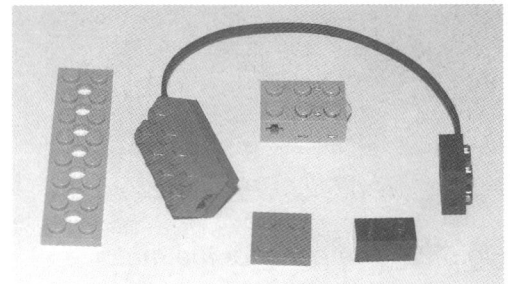

Fig.5.20 The parts needed to complete the front wheel assembly

Fig.5.21 The rear of the assembly

Fig.5.22 The front wheel assembly fitted to the chassis

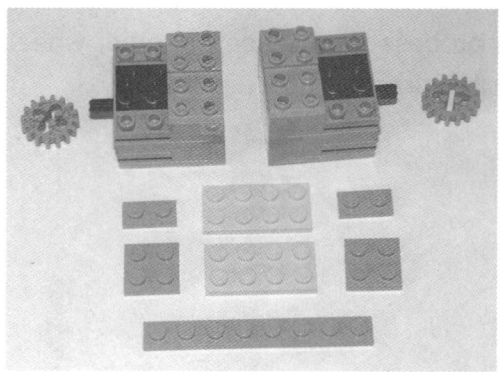

Fig.5.23 The parts for the motor assembly

should just about be sufficient space for it between the two axles. Then fit the second 8 x 2 plate on top of the light sensor and the four 2 x 1 beams. The 2 x 2 plate is then fitted on top of the light sensor, and the touch sensor is added on top of this. The front of the touch sensor should be flush with the front of the light sensor. Finally, fit a 2 x 1 beam immediately to the rear of the touch sensor. The rear view of the completed assembly shown in Figure 5.21 should help to get everything in the right place. The complete assembly is fitted on the underside of the chassis, as far forward as possible (Figure 5.22)

Step 5 (Figures 5.23 - 5.25)

Now the motor assembly is constructed and fitted to the chassis. The parts required are shown in Figure 5.23. Start by positioning the two motors back to back, and then fit the two 4 x 2 plates on top of the motors to join them together. Fit the other five plates on the underside of the motors (Figure 5.24) and then add the two 16-tooth gearwheels on to the shafts of the motors. The motor assembly is then complete and ready to be installed at the rear of the chassis (Figure 5.25).

Fig.5.24 The underside of the motor assembly

Fig.5.25 The completed motor assembly installed on the rear section of the chassis

Step 6 (Figures 5.26 and 5.27)

Next the bumper assembly is constructed. The parts required are shown in Figure 5.26 and the finished unit appears in Figure 5.27. Start by putting the

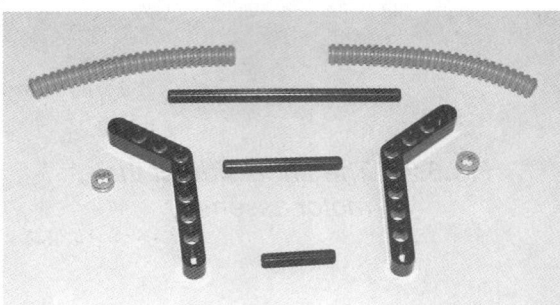

Fig.5.26 The components needed to construct the bumpers

smallest size of pulley (which acts here as a narrow fixing "nut") on the end of an axle about 39 millimetres long. Next fit the two black angled girder pieces onto the axle, side by side, with the axle going through the holes at the points in the girder pieces where the angle occurs. Then fit another small pulley at the opposite end of the axle to the first one. The girder pieces have holes to take axles at the very front and rear. The short axle (about 23 millimetres long) is placed in

Fig.5.27 The completed bumper is not fitted until the RCX unit is added to the chassis

the holes at the rear (i.e. at the end of the longer section). The largest of the three axles is about 79 millimetres in length, and it fits through the holes at the front of the assembly. The two girder pieces must be as far apart as the first axle and the two small pulleys allow, and the longest axle protrudes by roughly equal amounts on each side of the assembly. To complete the unit the two flexible tubes, which act as the actual bumpers, are fitted onto the ends of the longest axle.

Step 7 (Figures 5.28 and 5.29)

The RCX unit is now added onto the chassis, making sure that the end with the windows for the infrared communicator is at the front. Matters are complicated somewhat by the bumper assembly, which must be positioned on the chassis before the RCX unit is fitted onto the chassis

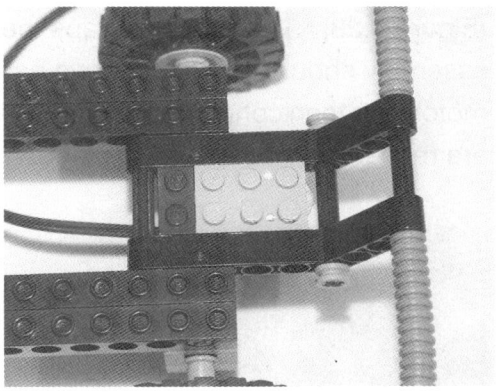

Fig.5.28 Fitting the bumper in place

Fig.5.29 The bumpers and RCX unit fitted to the chassis

(Figure 5.28). The RCX unit traps the bumper assembly in position, and the assembly should be free to move backwards and forwards slightly. The two motors are then connected to outputs A and C, and the touch and light sensors are respectively connected to inputs 1 and 2 (Figure 5.29).

Fig.5.30 The guard to keep the lead away from the front wheels

Step 8 (Figures 5.30 and 5.31)

It is advisable to add a couple of 2 x 1 beams (Figure 5.30) on the underside of the chassis to keep the lead from the light sensor away from the wheels. Otherwise this lead might occasionally foul the wheels and impede the performance of the robot. You should then have the completed robot of Figure 5.31.

Software

A number of components are required for the Visual BASIC program that accompanies Feedbackbot. They are all listed here with notes on positioning any setting adjustments required for each one. Figure 5.32 shows the program in operation, and should make it easier to copy the screen layout.

Spirit.OCX

Make the icon small and place it somewhere out of the way, such as the top left-hand corner of the screen.

Timer1

Make the icon small and place it somewhere out of the way, such as in the top right-hand corner of the screen. Set the interval property to 20.

Fig.5.31 The completed robot, ready for testing

Line1

There are various ways that light readings can be displayed onscreen, but for this application it is essential to use a method that enables you to quickly see and interpret any changes in level. The method used here is to have a vertical line on the screen, with the height of the line lengthening and shortening as the sensed light level goes up and down. This type of display is known as a bargraph incidentally, and is similar to the audio level indicators often fitted to hi-fi equipment. You can not mistake the line icon in the toolbox for anything else. It is simply a diagonal line. Draw a line anywhere on the screen and then use the Properties window to give it these co-ordinates.

X1   1000

X2   1000

Y1   4500

Y2   1000

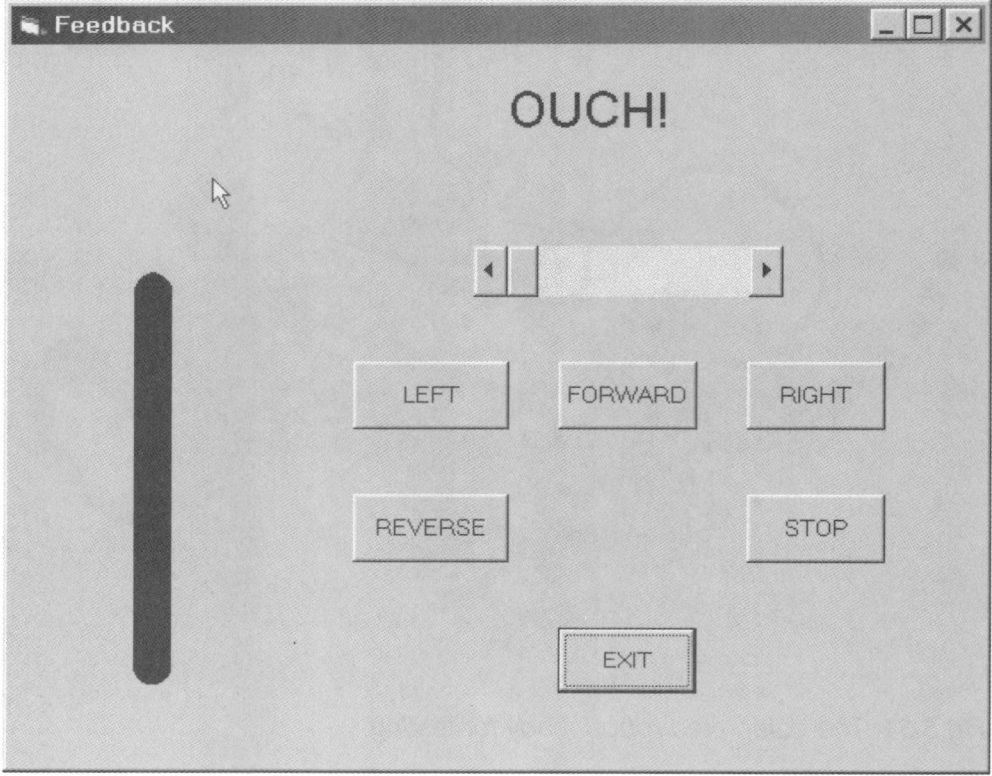

Fig.5.32 The completed program for Feedbackbot in operation

Also set the BorderWidth property to a higher figure than the default value so that the line is made broader and easier to see. A value of around 20 to 25 is about right.

Label1

Place this very near the top of the screen and offset to the right of centre, and either set it to auto-size or make it fairly large. Set a larger font size of about 18. Its function is to provide a warning if Feedbackbot runs into something.

HSrollBar1

Position this below Label1, and make it reasonably large so that it is easy to operate. The Max setting should be changed to 7 in the Properties window. It is used as a power control, and can be labelled accordingly if desired.

Command1

This control button is positioned well towards the bottom of the screen and a little to the right of centre. Its caption is changed to EXIT.

Command2

The first in a row of three buttons below the speed control, its caption is changed to LEFT.

Command3

The third in the row of three control buttons, the caption of this one is changed to RIGHT.

Command4

The second in the row of three command buttons, the caption of this one is changed to FORWARD.

Command5

This button is placed below Command3 (RIGHT) and its caption is changed to STOP.

Command6

This command button is positioned below Command2 (LEFT), and its caption is changed to REVERSE.

Listing

With the form design complete, double-click on the form to bring up the Code window, and then add the necessary lines of code so that the program matches the listing provided here:

```
Private Sub Command1_Click()
Spirit1.CloseComm
End
End Sub
```

```
Private Sub Command2_Click()
Spirit1.On "02"
Spirit1.SetFwd "2"
Spirit1.SetRwd "0"
End Sub

Private Sub Command3_Click()
Spirit1.On "02"
Spirit1.SetFwd "0"
Spirit1.SetRwd "2"
End Sub

Private Sub Command4_Click()
Spirit1.On "02"
Spirit1.SetFwd "02"
End Sub

Private Sub Command5_Click()
Spirit1.Off "02"
End Sub

Private Sub Command6_Click()
Spirit1.On "02"
Spirit1.SetRwd "02"
End Sub


Private Sub Form_Load()
Spirit1.InitComm
Spirit1.PBTxPower 1
End Sub

Private Sub Timer1_Timer()
Spirit1.SetSensorType 0, 3
Spirit1.SetSensorMode 0, 4, 0
Spirit1.SetSensorType 1, 1
```

```
Spirit1.SetSensorMode 1, 1, 0
If Spirit1.Poll(9, 1) = 1 Then Label1.Caption = "OUCH!" Else:
Label1.Caption = ""
barheight = Spirit1.Poll(9, 0)
barheight = barheight * 40
barheight = barheight + 500
barheight = 5500 - barheight
Line1.Y2 = barheight
Spirit1.SetPower "02", 2, HScroll1.Value
End Sub
```

Much of this program is the same as the remote control program described previously, so here we will only deal with the new features. There are two of these, which deal with reading the touch sensor and the light type. The sensors are handled by the Timer1 subroutine. The part of this routine that deals with the touch sensor is the simpler of the two. The touch sensor is connected to input 2, and the second SetSensorType and SetSensorMode commands set input 2 to switch type operation in the Boolean mode. This means that the value read from input 2 is 0 when the sensor is not activated or 1 when it is.

Most of the methods of reading a sensor are only applicable to indirect operation, and do not work at all in the immediate mode. However, the Poll command does provide a means for the PC to read sensors. Once read from a sensor, a value can be used in much the same way that it would be used in a program running in the RCX unit. It can be used in a conditional instruction such as an If command, or it can be placed in a variable. Note though, that it is Visual BASIC conditional instructions and variables that we are concerned with here, and that the RCX versions of these are not available when controlling a robot from the PC in the immediate mode.

The If command reads the touch sensor and displays a suitable message on the screen when the robot rams something. This instruction tests to see if the value read from the sensor is 1, and an If...Then...Else structure prints a warning message on the screen if it is, or a blank caption if it is not. In the event of a collision "OUCH!" is printed on the screen, but you can obviously replace this with anything you like. The Poll command has two parameters, and the first of these is a number from 0 to 16 that selects the source. The second defines the source more precisely, such as giving the number of a variable or input. Note though, that this system does not operate in exactly the same way as the source parameter for a downloadable command. Some of the available parameters do not apply to the RCX unit or are unused. The Software Developer Kit gives a full list of sources, but here are the ones you are most likely to use:

Number	Source selected	Second parameter
0	Variable	Variable's number (0 - 31)
1	Timer	Timer's number (0 - 3)
9	Sensor value	Input number (0 - 2)
12	Raw sensor value	Input number (0 - 2)
13	Boolean sensor value	Input number (0 -2)

There are actually three source numbers that select an input, and the one chosen determines the way in which the data is presented. With a source value of 12 for example, the value read is the raw value from the analogue to digital converter, giving a value in the range 0 to 1023. Here we wish to read the value from the sensor in the form it would be shown on the RCX unit's display (i.e. 0 or 1), and a value of 9 is therefore used. Like some of the downloadable commands, inputs 1 to 3 are selected using values of 0 to 2, so a value of 1 is used to select input 2. Note that the Poll command is strictly for use in the immediate mode, and that it can not be used as a downloadable instruction.

The first set of SetSensorMode and SetSensorType commands set up input 1 to read the light sensor. Readings are given as a percentage (0 to 100), as is normal when reading a light sensor. Another Poll command is used to read the light sensor, and the returned value is placed in the variable called "barheight". This variable is used as the Y2 co-ordinate for Line1, and it therefore controls the vertical position for the top of the line. Some mathematical manipulation is required first though.

The range of values read from the sensor is rather limited when converted into screen co-ordinates, which would give a short line with little change in its height. The returned value is multiplied by 40 in order to give more usable scaling. Things are complicated by the fact that the co-ordinate system has 0 at the top, which would give a line that grew from the top downwards. The bargraph display would be usable in this form, but probably most people would prefer a conventional type that grows from the bottom upwards. There is a further complication in that we require the line to start a little way up the screen, and not right at the bottom. Adding 500 to the returned value produces the offset from the bottom of the screen. The inversion of the line is obtained by deducting this value from 5500. It may seem that the mathematics is slightly out here, but the calculations have been adjusted slightly to take into account that the actual range of values from the light sensor is around 30 to 100, and not the nominal 0 to 100. The values used in the listing should give good results in practice.

Power setting

For this type of system to function properly it is essential that the infrared transmitter at the PC and the one in the RCX unit are both set for high power (long range) operation. The infrared transmitted connected to the PC's serial port is set to high power operation via the switch at the front of the unit. The RCX unit can only have its power setting changed by a command from the PC. It will probably default to long range operation, but it is as well to make sure by sending a suitable command when the program first starts up. This is the purpose of the PBTxPower command attributed to the form. There are two power settings available, which are 0 for low power and 1 for long range operation.

In Use

During the NASA moon landings a camera on the moon was sometimes guided by someone on planet Earth. This proved to be a very difficult operation due to the delay of a second or so as the pictures came back to Earth, followed by another delay of over a second as any control signals were sent to the camera. The situation with Feedbackbot is nothing like as bad as this, but there is a delay of a fraction of a second when a control message is sent, and another small delay when any feedback is sent from the robot. You have to learn to compensate for this when driving the robot, and it may take a little practice before you get the hang of it. It is advisable to use a low power setting for the motors so that the robot moves reasonably slowly, maximising your chances of staying fully in control of it.

A good way to test Feedbackbot is to place a torch in the corner of a room and then try to ram the torch without looking at the robot. Instead, rely on the feedback to guide you and indicate when a hit has been scored. If you get good at driving Feedbackbot you should be able to home in properly on any one of three or four torches place about half a metre apart.

Into practice

Moneybot

There have been many efforts over the years to make saving money more fun, and this has lead to the production of some ingenious moneyboxes. You have probably seen the sort of thing I mean, where you put some money in the box and things start to happen, such as tunes playing, lights flashing, etc. One of our local shops had something of this ilk that was used to raise money to support the RNLI. When you fed it with a coin a model lifeboat was launched down a slipway and into the sea. Unfortunately, this ingenious gadget has a design flaw. It was not screwed to the counter and some uncharitable soul stole it!

This moneybox design is not in the same class as the lifeboat moneybox, but it should amuse your family and friends. The idea is to have a chute that is almost horizontal, but is aimed up in the air slightly. If a coin is placed on the chute, the chute automatically lowers and the coin runs down the chute and into a box. The chute then returns to its original position in readiness for another coin. There is a twist to things in that the unit will only perform if it likes the colour of your money. Actually it will work just as well with gold, silver, or copper coloured coins, but it will only deposit coins in the box that are reasonably new and shiny. If you feed it with an old and dull coin it will not perform.

The completed moneybox is shown in Figure 6.1. If you require the ultimate in kitsch you can put together a Lego box to catch the money, such as the one shown in Figure 6.1. I would definitely not recommend using the unit to drop coins onto an expensive or delicate moneybox. The unit is basically just

Fig.6.1 Moneybot, complete with Lego moneybox!

an arm mounted on a tower, with one motor controlling the angle of the arm via a worm drive. A step-down gear plus the innate step-down provided by the worm drive ensures that everything happens in a slow and well-controlled fashion.

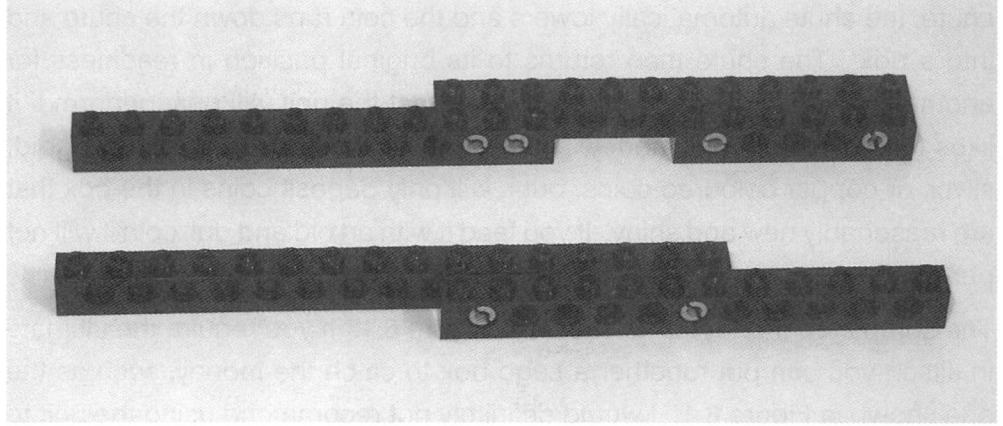

Fig.6.2 The completed side sections

Step 1 (Figures 6.2 - 6.5)

Construction starts with the base section. Two side pieces are required, and one of these consists of a 12 x 1 beam connected to a 16 x 1 beam using two pegs. The other consists of two 12 x 1 beams pegged together, and a 6 x 1 beam pegged onto these. Both completed side sections are shown in Figure 6.2. The base section is completed using the parts shown in Figure 6.3. Start by adding the five plates under the side sections, as shown in Figure 6.4. Then add the four beams on the top side of the assembly (Figure 6.5).

Fig.6.3 Additional parts for the base

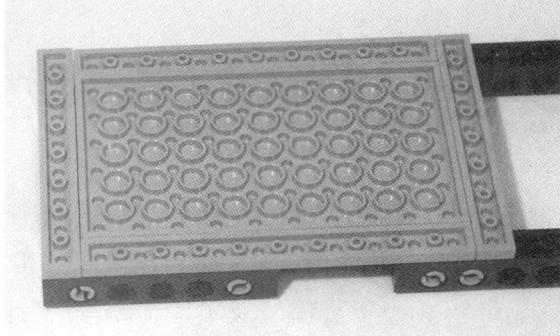

Fig.6.4 The plates fitted on the base

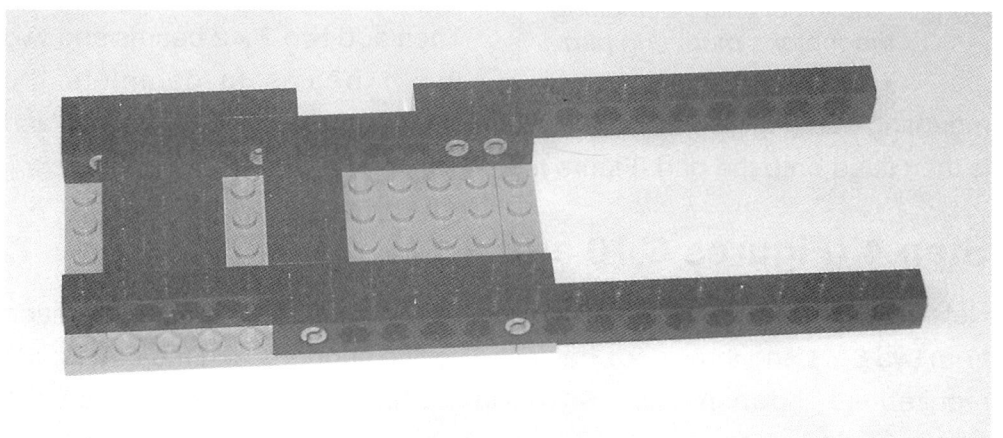

Fig.6.5 The finished base section

Fig.6.6 Strengthening the base

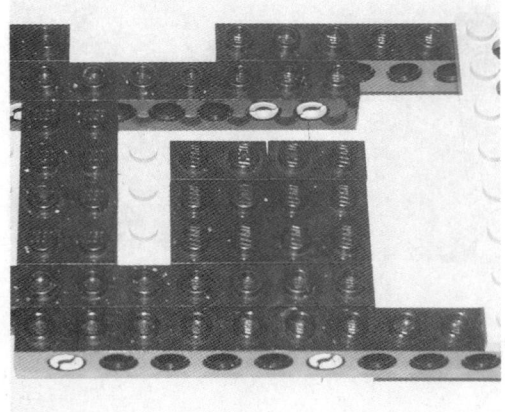

*Fig.6.7 The initial stage of making
the motor's mounting pad*

Step 2 (Figure 6.6)

The open end of the base requires some strengthening, and it is also fitted with a pad that will eventually take the RCX unit. First fit an 8 x 2 plate at the end of the base section, on the underside. A 6 x 1 beam is then fitted on this plate, one unit in from the end of the base section. The pad for the RCX unit, which is made from two 8 x 2 and one 10 x 2 plate, is then fitted on the top of the base section.

Step 3 (Figures 6.7 - 6.9)

The motor is mounted on the base section, and it requires a mounting pad made from beams. Start by fitting a 4 x 2 beam and two 2 x 1 beams onto the base (Figure 6.7). Then add two 2 x 2 beams and two 2 x 1 beams to complete the mounting pad (Figure 6.8). The motor is fitted with an 8-tooth gearwheel and is then fitted onto the pad (Figure 6.9).

Step 4 (Figures 6.10 and 6.11)

Next the worm drive is added. This requires two supports that are each made from two 6 x 1 and three 4 x 1 beams. The two supports and the other parts required are shown in Figure 6.10. Having built the two supports from the beams fit a fixing "nut" onto the end of an axle about 47 millimetres long.

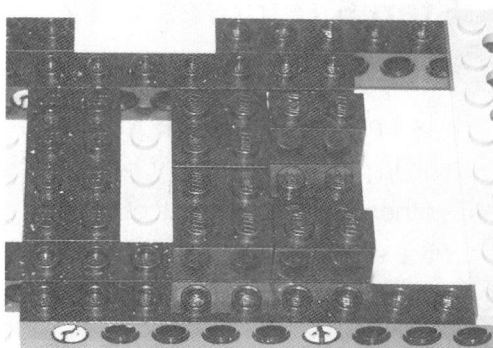

Fig.6.8 *Adding some more beams completes the mounting pad for the motor*

Fig.6.9 *The motor in position*

Fig.6.10 *The parts for the worm drive, including two assembled supports*

Fig.6.11 *The completed worm drive*

Then fit the axle through the top middle hole in a support, add the worm gear onto the axle, fit the axle through the top middle hole in the other support, and then add a 40-tooth gearwheel. The completed assembly is then added onto the base section, making sure that the 40-tooth gearwheel meshes properly with the 8-tooth type on the motor (Figure 6.11).

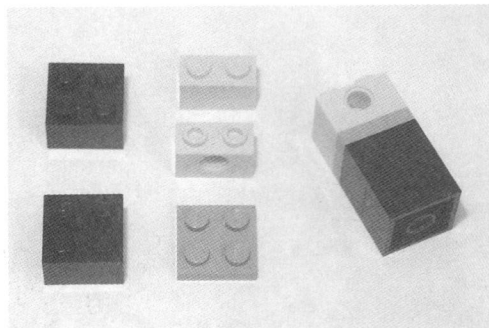

Fig.6.12 The parts for a support and a completed support

Fig.6.13 Both supports in position on the base

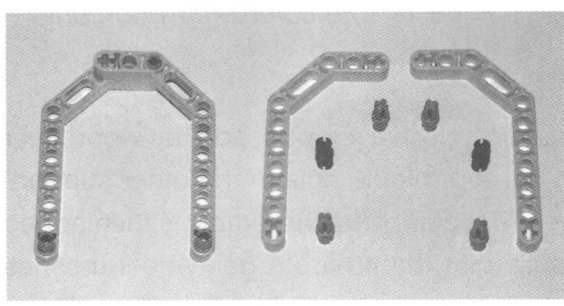

Fig.6.14 The towers are very simple

Step 5 (Figures 6.12 and 6.13)

It is important that the towers are held in place quite rigidly, since there is otherwise a strong risk of the worm drive slipping. Two mini towers are therefore used to add support to the main towers. The parts for one of the reinforcing blocks are shown in Figure 6.12, together with an assembled block. This assembly is very simple, and it is just a matter of adding two 2 x 2 blocks on top of each other, adding a 2 x 2 plate on top of this, and then adding two 2 x 1 beams on the plate. Note that one of these beams must be a type having a hole so that it can be pegged to a tower. The completed blocks are fitted onto the base section (Figure 6.13), making sure that the hole at the top of each block is facing outwards so that it can be pegged to a tower.

Step 6 (Figures 6.14 and 6.15)

To complete the base section the two towers are fitted onto the base. Figure 6.14 shows the parts needed for one tower, together with a completed tower. The two double angled girder pieces are joined together at the top using two pegs. These are pegs that have the standard fitting at one end and an axle style fitting at the other. Two pegs of the same type are used to fix the bottom of each tower to the base section. A black peg is fitted five holes up on one side of each tower, and these two pegs fix the towers to the two reinforcing sections on the base. Note that standard grey pegs will fit here, but the black pegs fix the towers in place much more firmly. This is important,

Fig.6.15 The finished towers in position

because more than a small amount of play in the towers could result in the worm drive losing its grip. Incidentally, the two towers are identical and are not "mirror images".

Step 7 (Figures 6.16 - 6.18)

Now we start construction of the arm. The parts required to build one side of the arm are shown in Figure 6.16. Start by fitting a 6 x 1 plate and a 10 x 1 plate on top of the 16 x 1 beam. Then add the other 10 x 1 and 6 x 1 plates on top of these. The shorter section of the angled girder piece is then fixed to the beam using two "rivets". The 7-unit long half-width girder piece is then "riveted" to the other side of the beam. Note that you

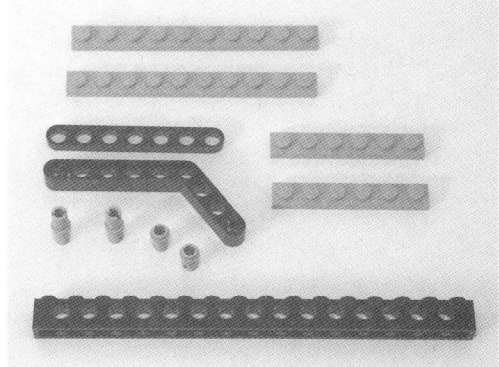

Fig.6.16 The parts for one side of the arm

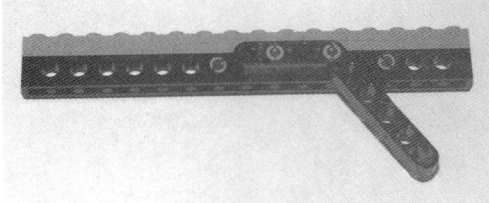

Fig.6.17 The completed side section

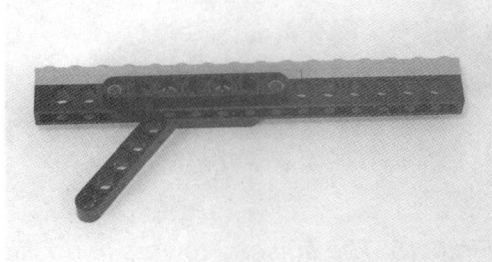

Fig.6.18 The coin guide can be seen in this view

must use the pegs that have the standard fitting at one end and the short fitting at the other. The short fitting goes into the half-width girder piece and should not protrude from it. The standard pegs would protrude from the girder and would prevent the coin from fitting properly into the arm. The purpose of the half-width girder piece is to narrow the slit for the coin so that it has to be reasonably upright. This gives good reflectivity back to the light sensor.

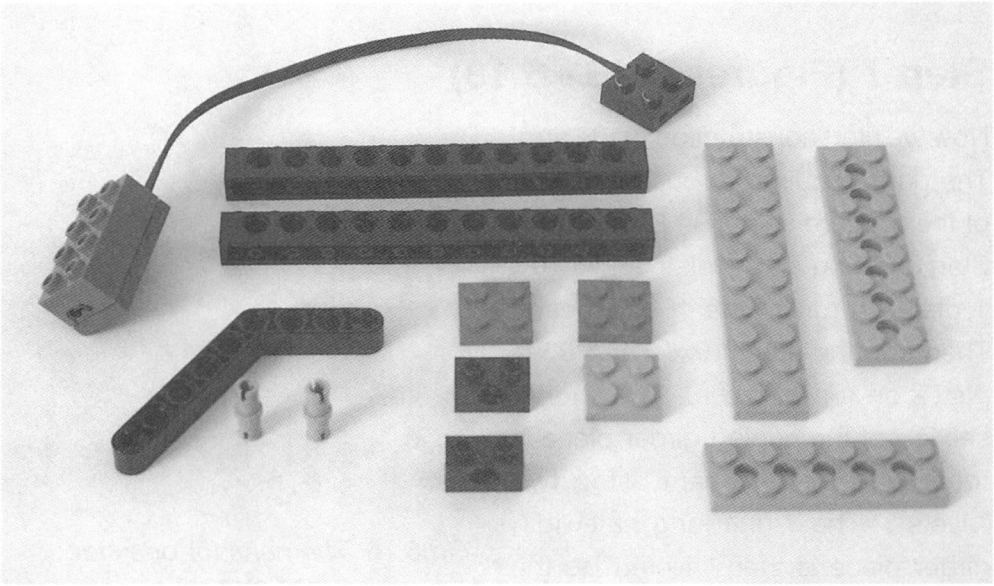

Fig.6.19 The parts required for the other side of the arm

Step 8 (Figures 6.19 - 6.21)

The other side of the arm is a little more complex since it must incorporate the light sensor. The parts required are shown in Figure 6.19. The two 12 x 1 beams are joined side by side by fitting an 8 x 1 beam on top of the beams. The plate is offset from the middle of the beams to leave a 4 x 2 area at one end that is fitted with two 2 x 2 plates. The two 2 x 1 beams are then placed side by side and joined using a 2 x 2 plate fitted on top. The sensor and the 2 x 1 beams are then joined with

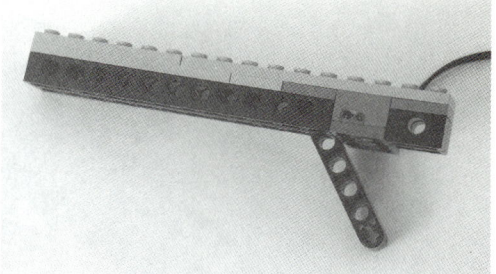

Fig.6.20 The finished side piece

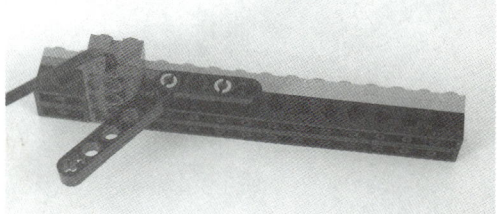

Fig.6.21 The other side of the assembly

the 12 x 1 beams using a 10 x 2 plate fitted on top of them. This leaves a 6 x 2 area on top of the beams, which is fitted with a 6 x 2 plate. To complete this side section the angled girder piece is pegged in place.

Step 9 (Figures 6.22 - 6.24)

The two sides are fixed together using eight 4 x 2 plates fitted on their undersides. A 2 x 1 beam is fitted at the rear of the assembly between the two side sections, and then a sloping 2 x 1 beam is placed on top of that. Figure 6.22 shows the pieces required. Figures 6.23 and 6.24 show the top and bottom sides of the completed arm assembly.

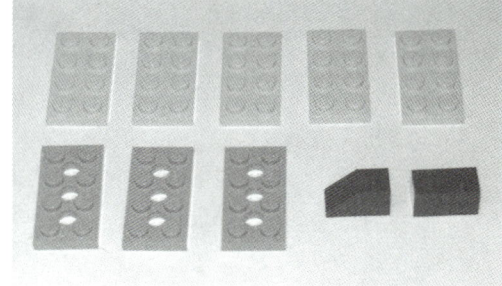

Fig.6.22 The parts required to join the two sides of the arm

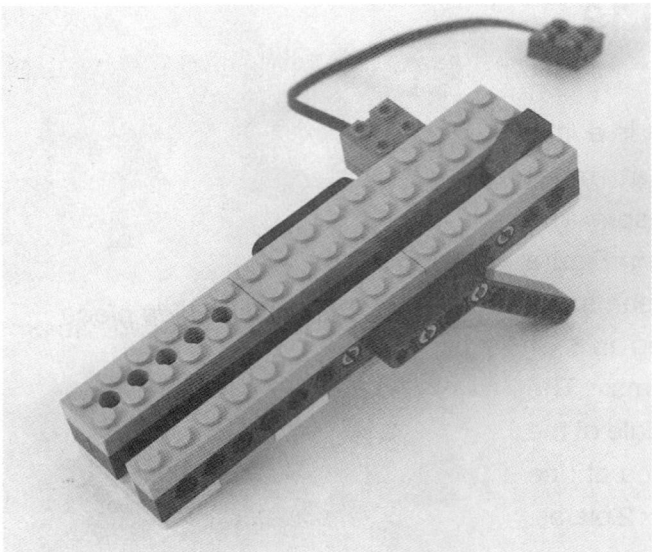

Fig.6.23 The top view of the completed arm, which is now ready to be fitted on the towers

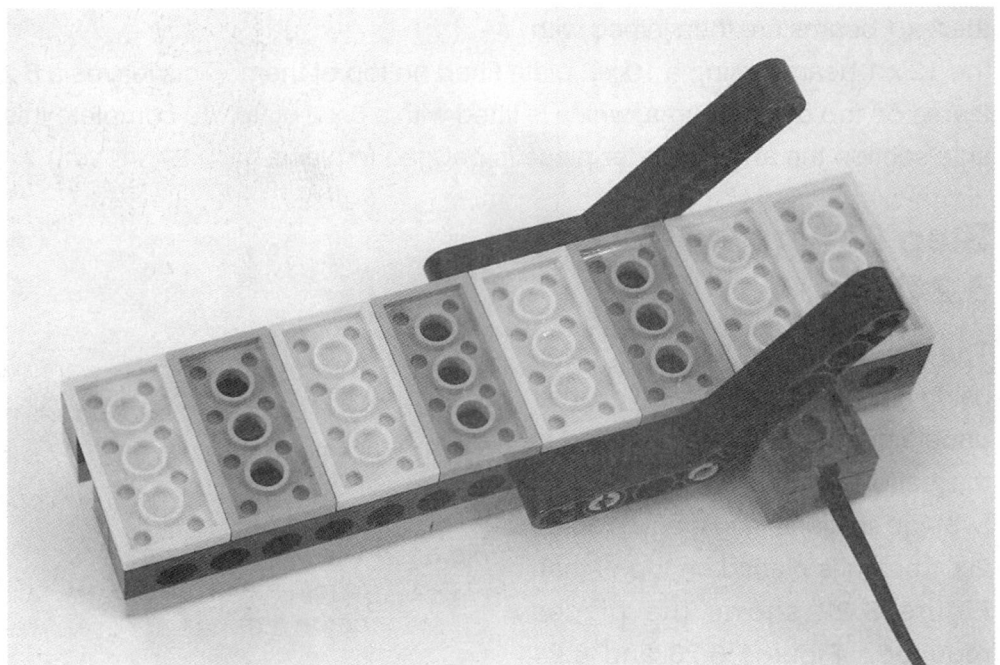

Fig.6.24 The underside view of the finished arm

Step 10 (Figures 6.25 - 6.27)

The completed arm is now mounted on the tower and coupled to the worm gear. The parts required are shown in Figure 6.25, and Figure 6.26 shows a close-up of the parts in position. With the arm assembly held in position, thread an axle about 95 millimetres long into one hole at the top of the tower and then into one side of the arm assembly. A fixing "nut", a 24-tooth gearwheel, and another "nut" are then added to the axle, after which it is threaded through the other sides of the arm assembly and the tower. A "nut" is then added at each end of the axle and the other two "nuts" are moved as far

Fig.6.25 *The parts used to mount the arm*

Fig.6.26 *The arm fitted in place*

outward along the axle as possible. This should keep everything accurately in position. With a large number of items on the axle it can be difficult to manoeuvre the axle and the other parts into position. Take things slowly and carefully, moving one piece at a time if necessary. Try to force things and you will probably find yourself back with a kit of parts again. To complete this stage of construction, move the 24-tooth gearwheel so that it meshes with the worm drive correctly. Then slowly turn the 40-tooth gearwheel so that the arm is set just above a horizontal position (Figure 6.27).

Fig.6.27 The arm in the correct standby position

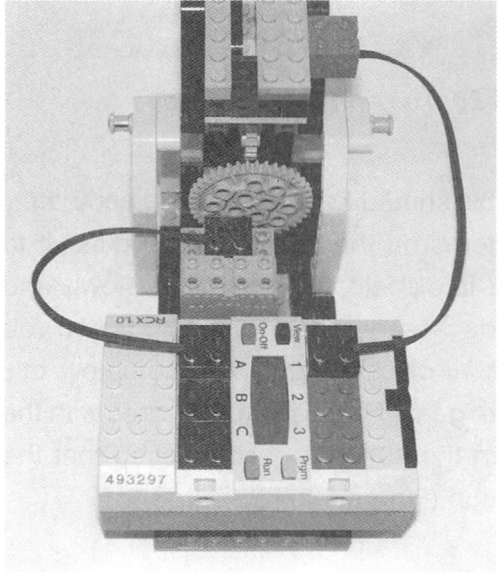

Step 11 (Figures 6.28 - 6.30)

To complete the unit fit the RCX unit onto its mounting pad on the base, connect the light sensor to input 1, and finally add a lead to connect the motor to output A (Figure 6.28). Figure 6.29 and 6.30 show the finished unit.

Fig.6.28 There are only two connections to the RCX unit

Figs.6.29 and 6.30 Two views of the completed Monybot robot

Moneybot RCX code

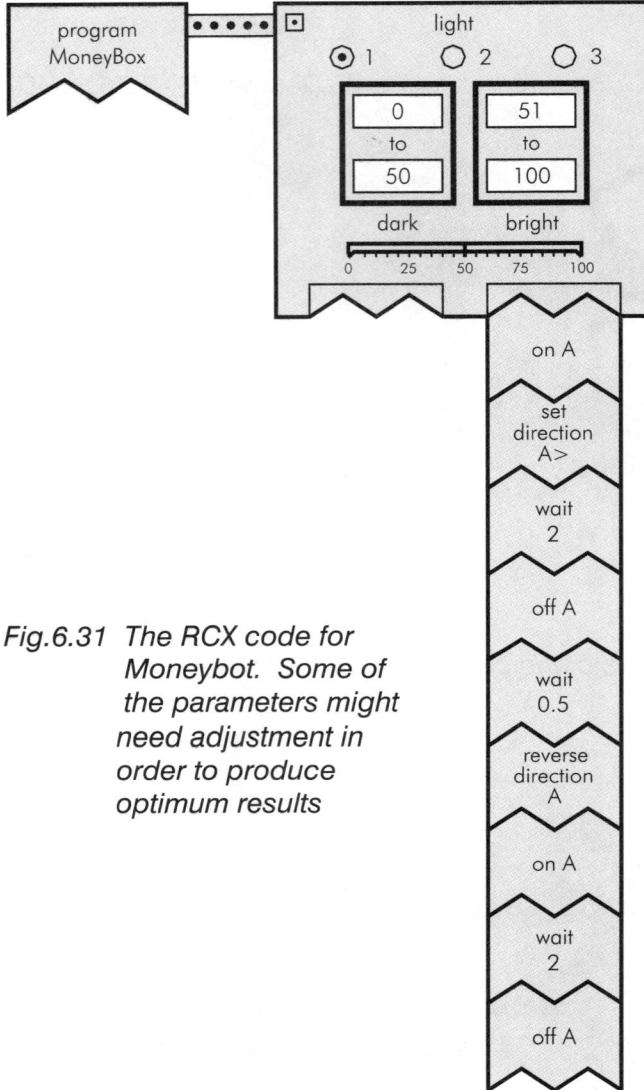

Fig.6.31 *The RCX code for Moneybot. Some of the parameters might need adjustment in order to produce optimum results*

The RCX code for the robot moneybox is shown in Figure 6.31. The program monitors the light sensor on input 1, and the list of commands is performed if the returned value is 51 or more. As usual with the light sensor, you might need to do some "fine tuning" of the threshold value in order to get things working really well. The threshold level is exceeded when a suitably bright coin is placed in the breech, and the series of commands is then commenced. First the motor is switched on and set to go forward.

The Wait command keeps the motor going forwards for two seconds, which should lower the angle of the arm sufficiently to get the coin rolling. The motor is then switched off, and another Wait command keeps it switched off for half a second. This gives time for the coin to get well clear of the arm. The direction of the motor is then reversed and it is switched on for a further two seconds. This returns the arm to its original position.

VB code

The Visual BASIC listing requires the usual form equipped with Spirit.OCX and two buttons captioned "DOWNLOAD" and "EXIT". The code window should then be edited to give this listing:

```
Private Sub Command1_Click()
With Spirit1
.InitComm
.SelectPrgm 3
.BeginOfTask 0
.SetSensorType 0, 3
.SetSensorMode 0, 4, 0
.Loop 2, 0
.If 9, 0, 0, 2, 50
.On "0"
.SetFwd "0"
.Wait 2, 200
.Off "0"
.Wait 2, 50
.AlterDir "0"
.On "0"
.Wait 2, 200
.Off "0"
.EndIf
.EndLoop
.EndOfTask
End With
End Sub

Private Sub Command2_Click()
Spirit1.CloseComm
End
End Sub
```

After the usual setting up, which includes two commands to set input 1 for percentage mode operation with a light sensor, the program goes into an infinite loop. The program is then a direct equivalent to the RCX version, with an If instruction checking to see if the threshold level has been exceeded. When it is, a list of commands is performed, and these have the same effect as their RCX code equivalents.

Improved VB listing

This version of the program incorporates one or two additions and refinements:

```
Private Sub Command1_Click()
With Spirit1
.InitComm
.SelectPrgm 1
.BeginOfTask 0
.SetSensorType 0, 3
.SetSensorMode 0, 4, 0
.SetVar 0, 9, 0
.Loop 2, 0
.SetVar 1, 9, 0
.SubVar 1, 0, 0
.If 0, 1, 0, 2, 5
.On "0"
.SetFwd "0"
.Wait 2, 200
.Off "0"
.Wait 2, 50
.AlterDir "0"
.On "0"
.Wait 2, 200
.Off "0"
.EndIf
.If 0, 1, 1, 2, -3
.PlayTone 100, 10
```

```
.EndIf
.EndLoop
.EndOfTask
End With
End Sub

Private Sub Command2_Click()
Spirit1.CloseComm
End
End Sub
```

As we have seen previously, there is a slight problem when dealing with the light sensor. The value returned for a given set of conditions will vary slightly from one sensor to another. Also, and perhaps of greater importance, the returned value is to some extent dependent on the ambient light level. In fact the returned value seems to be heavily dependent on the ambient light level. Even if you adjust the threshold level to suit the particular sensor you are using, a different ambient light level the next time the robot is used could prevent it from performing properly. When using Spirit.OCX there is an easy way around the problem, which is to measure the light level when the program is first run, and to then test for deviations from this level. In effect the software is still set to respond to a particular threshold level, but each time the program is run the threshold value is automatically adjusted to suit the particular sensor used and the ambient light level.

The light sensor is read prior to entering the loop, with the returned value being placed in variable 0. Next the program goes into an infinite loop, as before, but the If instruction is preceded by two instructions. The first of these reads the sensor again, and places the returned value in variable 1. The second instruction deducts the original reading (in variable 0) from this reading and places the result in variable 1. If there has been an increase in the light value due to reflected light from a coin, the value now stored in variable 1 will be a positive value of typically around 5 to 15. If a suitable coin is not present in the breech, the value in variable 1 will be zero or a small value of either sign.

The If instruction tests for a value of more than +5, and the following instructions are performed if this test produces a positive result. These instructions lower the arm and return it to the original position, as before. The value used in the If instruction determines how shiny or otherwise the coin must be in order to activate the unit. Increasing the value makes the unit respond to only the shiniest of coins. It is best not to use a value of much less than 5 as this could result in minor changes in the ambient light level triggering the unit into action.

This version of the software has an added refinement in the form of a simple sound effect. Another If instruction provides the effect that looks for a reduction in the light value by more than 3. A short burst of low frequency sound is produced if this condition is met. In other words, Moneybot does its best to blow a raspberry at you if you try to feed it with a dull and old coin! The light sensor is aimed at black plastic under standby conditions, and it might seem as though no coin would reflect less light than the plastic. However, the plastic may be black, but it is quite shiny as well. Anything with a dark and non-shiny surface will produce a reduction in the light reading.

Handybot

Moneybot has an arm, but it is of limited use since it can only be raised and lowered. It would be of more use if it had a hand plus greater mobility in the arm. With only two motors in the Robotics Invention System there is a limit to the range of movements and facilities that can be provided, but adding a simple hand makes a robot arm more interesting and boosts its play value. Handybot has what are really two fingers that can be used to grip small objects, rather than something that could be accurately described as a hand. However, a simple two-finger grip is sufficient to perform some simple robot tasks.

Handybot, like Moneybot, can be used to put a coin in a moneybox. With Handybot you first press a button to activate the unit, and then place the coin between the two mechanical fingers as they close. Having grasped the coin, Handybot lowers its arm, releases the coin into the box, and raises its arm again. This is a modern version of an idea that goes back to Victorian times, and probably earlier. The original units were purely mechanical of course, and it was usually the beak of a toy bird that grabbed and manipulated the coin.

The angled girder pieces are well suited to use as mechanical fingers. The fact that a coupling via gearwheels results in the driven shaft going in the opposite direction to the drive shaft is something that can be exploited in this application. In the simple arrangement shown in Figure 3.32, the fingers can be opened and closed by driving either shaft. Handybot's hand is based on this scheme of things.

This robot is to a large extent based on Moneybot. In fact the base section and towers are exactly the same, as is the way in which the arm is mounted on the towers. The only

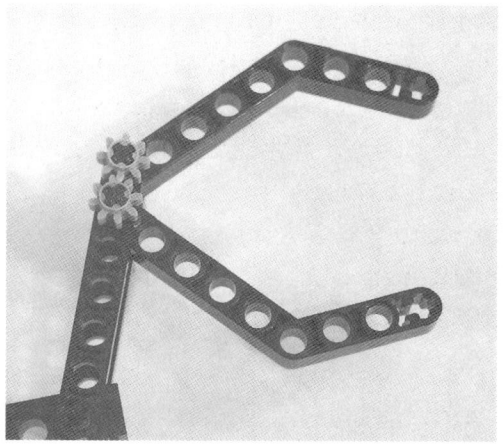

Fig.6.32 A simple mechanical hand

Fig.6.33 The additional plates

slight modification to the original base unit is the addition of a 6 x 2 plate and a 2 x 1 plate on top of the two supports for the towers (Figure 6.33). The 6 x 2 plate provides a mechanical link between the two supports that helps to keep everything as rigid as possible. This is important since Handybot's arm is rather heavier than that of Moneybot. The 2 x 1 plate is not essential, and is included mainly for cosmetic reasons. Another slight change is the addition of a touch sensor on top of the motor (Figure 6.34). Two views of Handybot are shown in Figures 6.35 and 6.36.

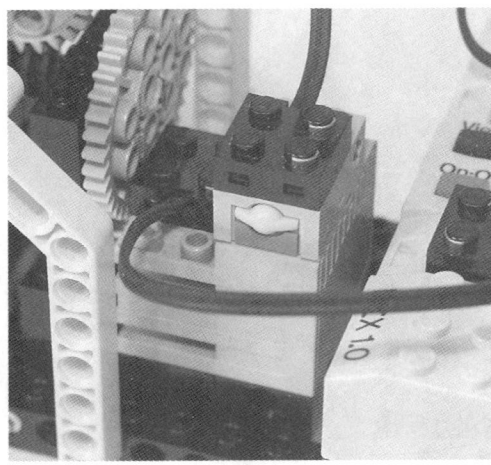

Fig.6.34 A touch sensor is added at any convenient place, such as on top of the motor

Fig.6.35 The completed Handybot, ready for action

Fig.6.36 The robot viewed from the other side

Step 1 (Figures 6.37 and 6.38)

First we require a basic arm to which the hand, motor, and drive system can be added. The parts required are shown in Figure 6.37, and the completed arm subassembly is shown in Figure 6.38. Three 12 x 1 beams are pegged

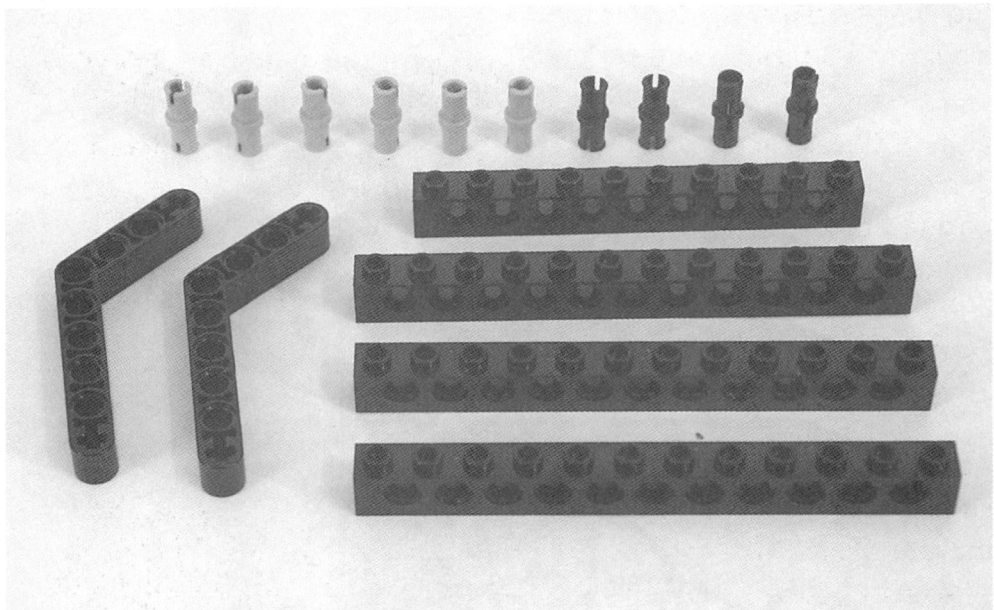

Fig.6.37 The parts required for the basic arm

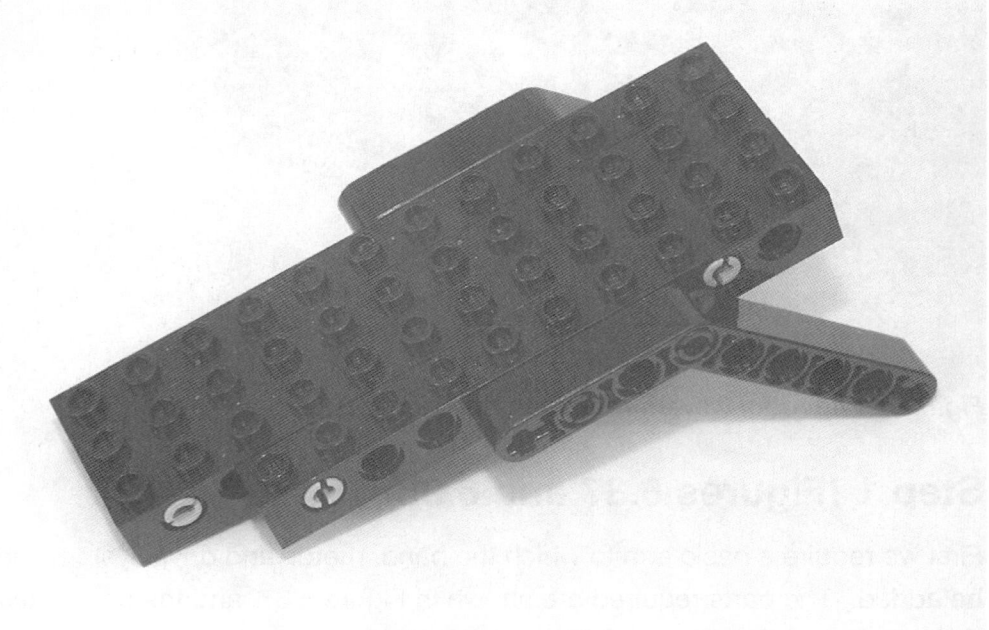

Fig.6.38 The completed basic arm, ready for the front section to be added

together to effectively form a single 12 x 3 beam. A 10 x 1 beam is then pegged onto the side of this assembly. The angled girder pieces are then pegged to the sides of this platform.

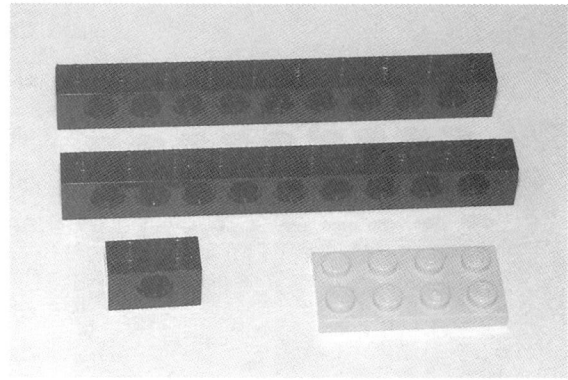

Fig.6.39 The additional arm pieces

Step 2 (Figures 6.39 and 6.40)

Adding the additional pieces shown in Figure 6.39 lengthens the arm slightly. The two 10 x 1 beams are fitted on top of the arm assembly, and they must have an overhang of 10 units. The 2 x 1 beam is then fitted, and the 4 x 2 plate is added on top of this and the other two beams (Figure 6.40).

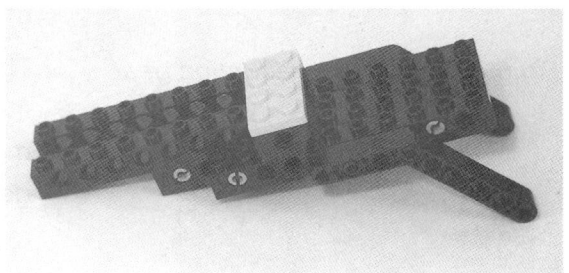

Fig.6.40 The finished arm assembly

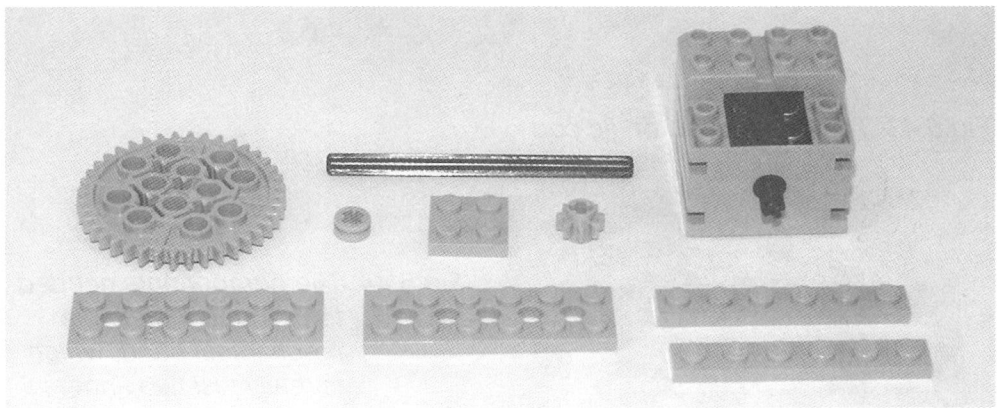

Fig.6.41 The motor and its associated parts

Fig.6.42 The 5 to 1 reduction gear

Fig.6.43 The motor fitted on its pad

Step 3 (Figures 6.41 - 6.43)

Now the motor and drive shaft are fitted to the arm. The parts required are shown in Figure 6.41. The two 6 x 2 plates fit on top of the rear section of the arm, as far forward as possible. The two 6 x 1 plates and the 2 x 2 type are then fitted on these to make a mounting pad for the motor.

The 8-tooth gearwheel is fitted onto the shaft of the motor, which is then mounted on its pad. Fit the 40-tooth gearwheel onto the end of the 63-millimetre axle, and then fit the axle into the arm so that the 40-tooth gearwheel meshes with the 8-tooth type on the motor. Finally, fit the smallest size of pulley wheel onto the axle to hold it in place.

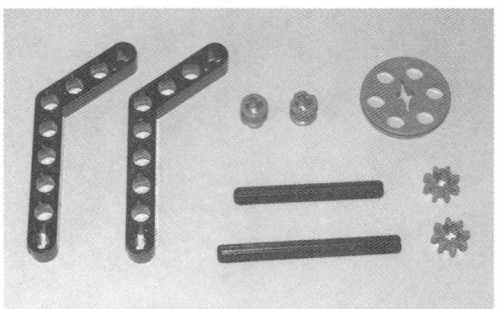

Fig.6.44 The components needed to produce the simple mechanical hand, which is really just two fingers

Step 4 (Figures 6.44 - 6.46)

The arm is now ready for the hand to be fitted at the end. The parts required are shown in Figure 6.44, and the two sides of the installed assembly are shown in Figures 6.45 and 6.46. Start by placing an 8-tooth gearwheel at the end of each axle. These are approximately 39 millimetres and 46 millimetres long. The girder piece that acts as the upper finger is placed in position between the end set of holes in the arm, and the shorter axle is threaded through the girder and the arm. The 8-tooth gearwheel should be on the same side of the arm as the 40-tooth gearwheel on the motor. A fixing "nut" is then added to keep the axle in place. Essentially the same process is then used for the lower finger, but once everything is in place the large pulley wheel is fitted onto

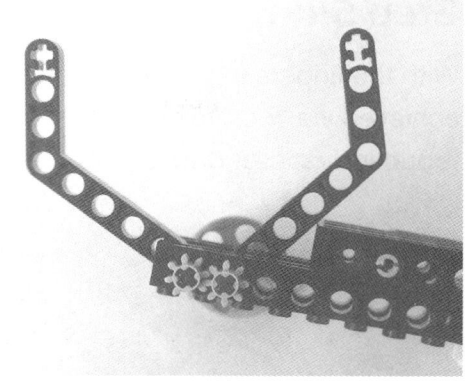

Fig.6.45 *Two gears link the fingers*

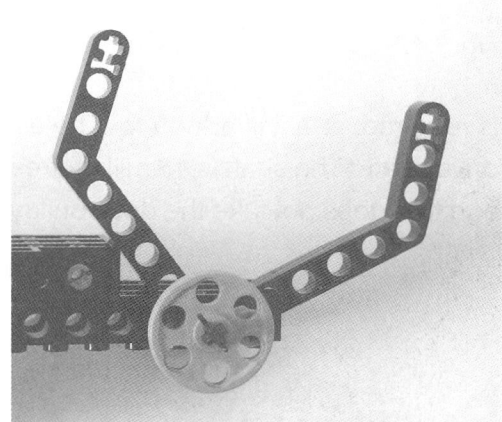

Fig.6.46 *The hand's drive pulley*

the very end of the longer axle. Set the fingers to the open position before fitting the second 8-tooth gearwheel, otherwise you may find that it will not fully open or that it will be set at an odd angle to the arm. It should be roughly perpendicular to the arm.

Step 5 (Figures 6.47 and 6.48)

With the fingers in place they must now be coupled to the motor, and this is achieved via two driving bands and two pairs of pulleys. There is already a reduction ratio of 5 to 1 from the motor to the existing drive shaft, and two further reductions of 3.5 to 1 give an overall ratio of over 60 to 1. This ensures that the fingers operate in a slow and controlled fashion rather than snapping shut and flying open. The parts needed for the two systems of driving bands are shown in Figure 6.47. Of course, two pulleys are already present on the arm assembly. An additional shaft is required, and it is added four holes further along the arm from the existing drive shaft. The shaft is 46 millimetres long, and a small pulley is used on the end that does not take the drive pulleys. This one simply stops the shaft from slipping out of place. Another small pulley is added at the other end of the shaft, followed by a pulley of the second largest size. The latter is fitted right at the end of the shaft. The small pulley is positioned so that it is aligned with the pulley on the finger mechanism. First fit a blue driving band to link the small pulley to the larger one on the finger mechanism. Adjust the fingers to the open position (their standby position) and then fit a black driving band to link the small pulley on the primary drive shaft to the larger pulley on the secondary shaft. This gives the completed drive mechanism shown in Figure 6.48.

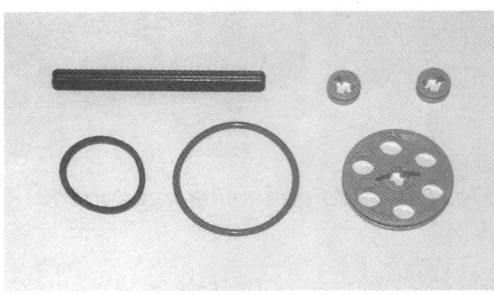

Fig.6.47 Parts for the drive system

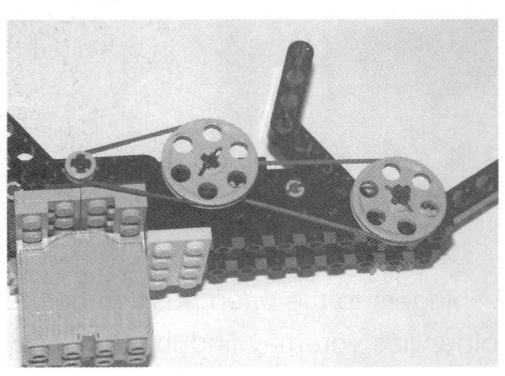

Fig.6.48 The finished drive system

Step 6 (Figures 6.49 and 6.50)

The arm is now more or less complete, but there is a slight problem in that the fingers are made from plastic that has a smooth and shiny finish. This makes it almost impossible for the fingers to grip something like a coin well enough to maintain a hold on it. The easiest and probably best

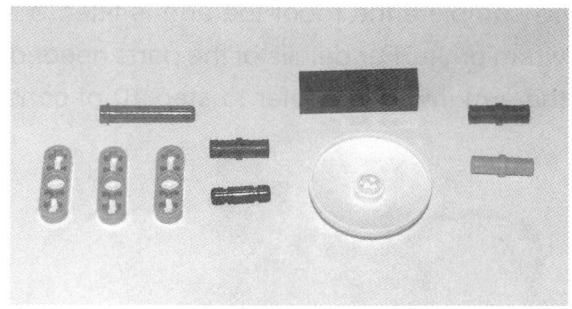

Fig.6.49 The additional parts for the hand

solution is to fit small pads of something like thin foam rubber material onto the "fingertips". Even something as basic as pieces of double-sided adhesive tape will do the job. However, once the pieces of tape are in position, dab your fingers onto them a few times to reduce the tackiness of the adhesive. Otherwise the robot will show a marked reluctance to part company with the coins, which is understandable but spoils the effect!

The other option is to try adding some pieces of Lego to give the fingers improved grip. The parts I used are shown in

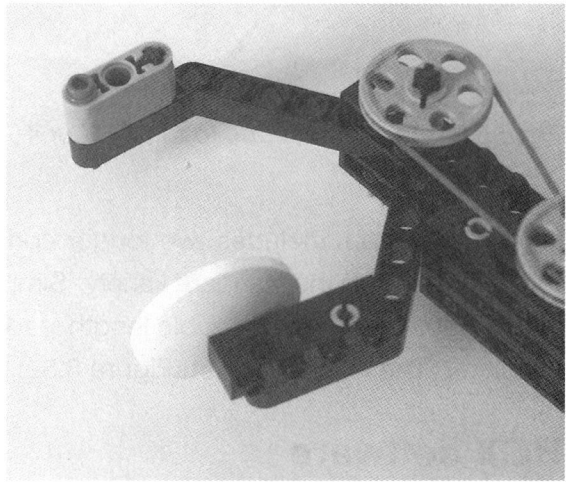

Fig.6.50 The completed hand

Figure 6.49, and the completed hand is shown in Figure 6.50. This modification works quite well, although the coins can tend to get stuck in the dish-shaped component on the lower finger, so some "fine tuning" of the design might be in order.

Step 7 (Figure 6.51)

To complete the robot the arm is fitted onto the towers and coupled to the worm drive. For details of the parts needed and the process of actually fitting the arm into place refer to step 10 of constructing Moneybot, as described

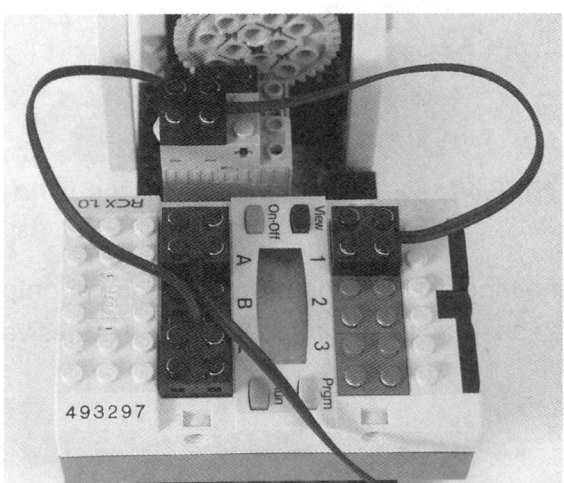

previously. The final stage is to connect the two motors and the touch sensor to the RCX unit. The motor that drives the arm up and down connects to output C, and the one that controls the mechanical fingers connects to output A. The touch sensor is connected to input 1. These connections are shown in Figure 6.51. Note that an ordinary connecting lead is too short to connect the RCX unit to the motor mounted on the arm. The Robotics

Fig.6.51 The connections to the RCX unit

Invention System includes two longer connecting leads, but these are very much longer than the standard variety. Simply connecting two standard length leads together to form a double length lead is a more practical solution to the problem. The finished robot (Figure 6.52) is then ready for testing.

RCX software

The RCX code for Handybot appears in Figure 6.53. A sensor watcher block monitors Input 1, and the list of commands is performed when the touch sensor is operated manually. First a "beep" sound is produced, to let you know that the switch has been activated successfully, and there is then a one second wait to give you time to get the coin in position. Both motors are set to forward operation, and then motor A is turned on for two seconds to close the fingers. This is followed by Motor C being turned on for three seconds to

Fig.6.52 The completed money grabbing robot, ready for testing

lower the arm. The direction of both motors is then reversed, and motors A and C are again turned on for two seconds and three seconds respectively. This opens the fingers and raises the arm back to its original position. The two and three second times will probably have to be "fine tuned" to get the arm and fingers operating in exactly the desired fashion. The previous programs in RCX code have used separate On and Wait instructions to turn on motors for a certain length of time. This keeps the RCX code and Visual BASIC versions as similar as possible. In this program On...For instructions have been used instead, which helps to prevent the program from growing an excessively long tail.

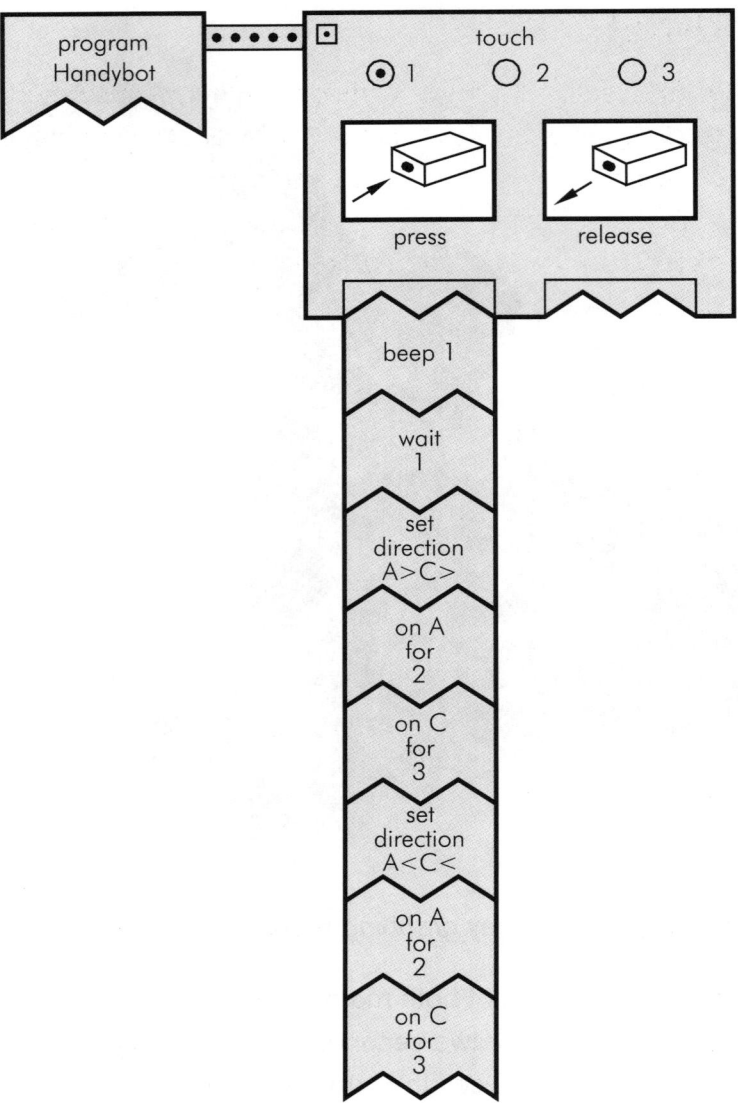

Fig.6.53 The RCX program for Handybot. Some of the times might benefit from some "fine tuning"

VB Handybot

The Visual BASIC version is very straightforward in operation and does not merit a detailed explanation. An If instruction in an infinite loop repeatedly tests the touch sensor until it is operated and a value of 1 is returned. The following list of instructions is then performed, and this provides the same sequence of events as the RCX code program.

```
Private Sub Command1_Click()
With Spirit1
.InitComm
.SelectPrgm 3
.BeginOfTask 0
.SetSensorType 0, 3
.SetSensorMode 0, 4, 0
.Loop 2, 0
.If 9, 0, 0, 2, 50
.On "0"
.SetFwd "0"
.Wait 2, 200
.Off "0"
.Wait 2, 50
.AlterDir "0"
.On "0"
.Wait 2, 200
.Off "0"
.EndIf
.EndLoop
.EndOfTask
End With
End Sub

Private Sub Command2_Click()
Spirit1.CloseComm
End
End Sub
```

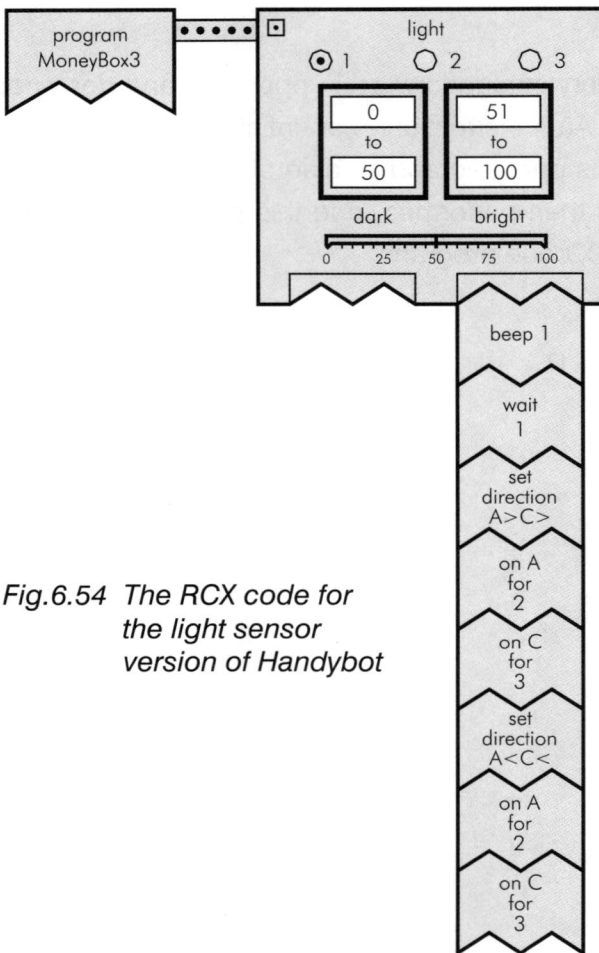

Light version

A light sensor can be used instead of the touch type, and you then place the coin in front of the sensor to get Handybot's approval. If the coin is shiny enough and Handybot "beeps" its "seal of approval", you then have one second to get the coin into position. The robot then goes through its standard routine. The version of the program in RCX code is provided in Figure 6.54. This is the same as the original apart from the fact that the sensor watcher is a light type. As usual, you might have to make some minor adjustment to the threshold level in order to

Fig.6.54 The RCX code for the light sensor version of Handybot

get the program working nicely. The higher the minimum light level setting, the shinier the coin must be in order to activate the unit.

This is the Visual BASIC program for the light sensor version of Handybot:

```
Private Sub Command1_Click()
With Spirit1
.InitComm
.SelectPrgm 1
```

```
.BeginOfTask 0
.SetSensorType 0, 3
.SetSensorMode 0, 4, 0
.SetVar 0, 9, 0
.Loop 2, 0
.SetVar 1, 9, 0
.SubVar 1, 0, 0
.If 0, 1, 0, 2, 5
.PlaySystemSound 1
.Wait 2, 100
.SetFwd "02"
.On "0"
.Wait 2, 200
.Off "0"
.On "2"
.Wait 2, 300
.Off "2"
.SetRwd "02"
.On "0"
.Wait 2, 200
.Off "0"
.On "2"
.Wait 2, 300
.Off "2"
.EndIf
.If 0, 1, 1, 2, -3
.PlayTone 100, 10
.EndIf
.EndLoop
.EndOfTask
End With
End Sub

Private Sub Command2_Click()
Spirit1.CloseComm
End
End Sub
```

This is much the same as the improved version of the Moneybot program, and merits little comment here. It has been modified to provide appropriate control of Handybot's two motors, and it also provides a "beep-beep" sound when successfully activated by a coin. In other respects it is the same as the original program.

Firm grip

Using pulley wheels and driving bands to transfer power to the mechanical hand has its advantages, one of which is that it is easy to couple a motor at one end of the arm to the fingers at the other end. On the face of it, the motor could simply be positioned at the same end of the arm as the fingers. In practice this does not work very well as it places a lot of weight at the top end of the arm. This makes life difficult for the motor that raises and lowers the arm, and generally makes the whole robot more unstable and unwieldy. A second big advantage of the pulley wheel system is that once the fingers have closed, the slippage in the drive band provides a simple clutch action. The fingers will tighten their grip but there is no danger of the robot doing itself any damage.

The main drawback of the pulley method is a lack of precision, with slippage and the elasticity of the driving band making things "a bit hit and miss". Also, the built-in clutch action keeps everything safe, but also limits the firmness of the robot's grip. For a more precise hand with a firmer grip it is necessary to resort to gears to drive the fingers, and this method is used in Gripperbot (Figure 6.55). The motor drives the main drive shaft via a 3 to 1 reduction gearing system, and the shaft drives the mechanical fingers by way of a worm drive. This gives a further reduction ratio of 24 to 1, producing an overall step-down of 72 to 1. The force in the fingers and the shafts driving them is potentially large enough to damage something, but the top of the arm is deliberately built in a slightly lightweight fashion. This ensures that no parts will be damaged if the motor attempts to drive the fingers beyond the fully open or closed positions. A few pieces will simply pop out of position, and they can simply be pressed back into place again.

Fig.6.55 Gripperbot uses a worm drive and gears to transfer power from the motor to the fingers

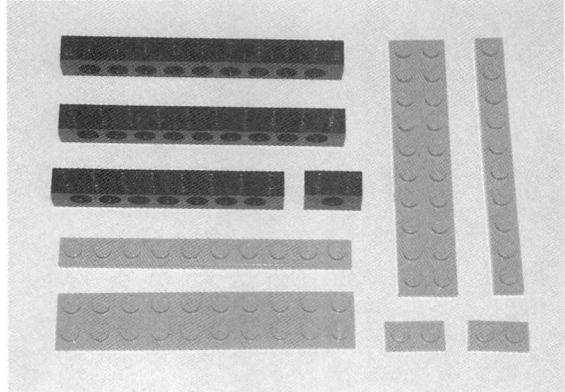

Fig.6.56 The parts for the platform

Fig.6.57 The beams added to the arm

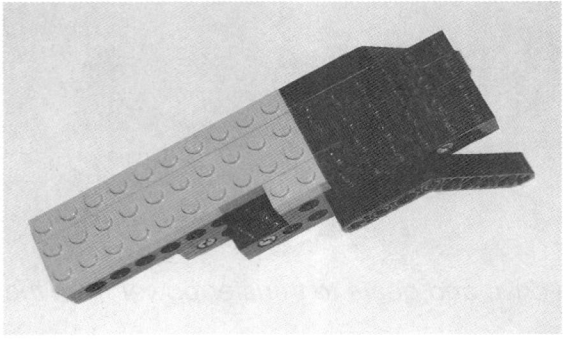

Fig.6.58 The finished platform in place

The base section used for Handybot is also used for Gripperbot. The basis of the arm is also the same, so complete step 1 of the Handybot building instructions and then follow these instructions:

Step 1 (Figures 6.56 - 6.58)

The arm is fitted with a raised platform at the front, and this requires the parts shown in Figure 6.56. Start by adding the two 10 x 1 beams, the 8 x 1 beam, and the 2 x 1 type to the basic arm assembly (Figure 6.57). The 10 x 1 beams overhang the front of the basic arm assembly by four units. Finally, the plates are added in two layers on top of the beams, but the 8 x 1 beam is not fully covered.

Fig.6.59 The parts for the hand

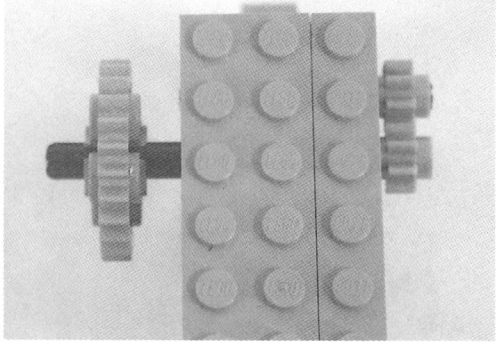

Fig.6.60 Rear view of the hand

Step 2 (Figures 6.59 - 6.61)

The finger mechanism is added next, and this is very similar to the one used in Handybot. The parts required are shown in Figure 6.59, and the finished assembly is shown in Figures 6.60 and 6.61. The two shafts are approximately 31 millimetres and 47 millimetres long.

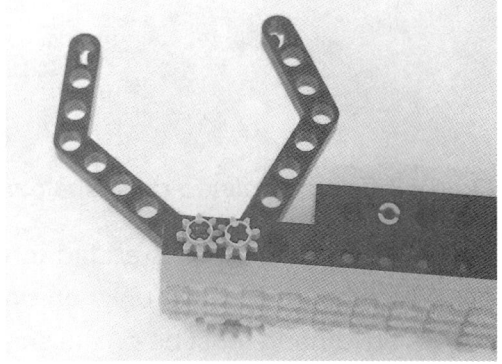

Fig.6.61 The completed hand in position

Step 3 (Figures 6.62 and 6.63)

The arm is now ready for the drive shaft to be fitted. Figure 6.62 shows the parts required, and Figure 6.63 shows the finished drive shaft assembly in position on the arm. The two 6 x 1 beams are added at the end of the arm. One is fitted at the

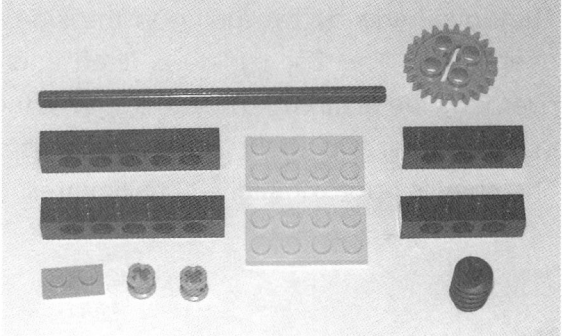

Fig.6.62 The drive system components

Fig.6.63 The completed drive shaft assembly fitted on the arm

very end, and the other is added three units down the arm. These beams overhang the arm by two units on one side and one unit on the other. The two-unit overhang must be on the same side as the 24-tooth gearwheel. The two 4 x 2 plates are fitted on top of these beams, leaving the area above the 24-tooth gearwheel uncovered. The two 4 x 1 beams are fitted across the other end of the platform on the arm, and there is no gap between them. They overhang by two units on the side of the arm that has the 24-tooth gearwheel. The 2 x 1 plate is fitted underneath these beams where they overhang the platform. The drive shaft is then fitted into the appropriate holes in the beams, adding the two gearwheels and two fixing nuts along the way. The shaft is approximately 95 millimetres long.

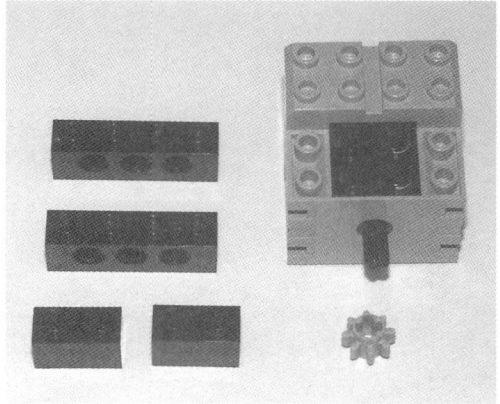

Fig.6.64 *The motor and associated parts*

Fig.6.65 *The motor mounting pad*

Step 4 (Figures 6.64 - 6.66)

To complete the arm the motor is fitted. The parts required are shown in Figure 6.64. The four beams are fitted to the rear of the arm to provide a mounting pad for the motor (Figure 6.65), the 8-tooth gearwheel is fitted onto the motor's shaft, and then the motor is mounted on its pad.

Fig.6.66 *The motor mounted on the pad*

Step 5 (Figure 6.67)

The arm is now ready to be fitted onto the towers. It is mounted in exactly the same way as the arms of the Moneybot and Handybot robots, and uses the same drive system. To complete the robot the motor on the base section is

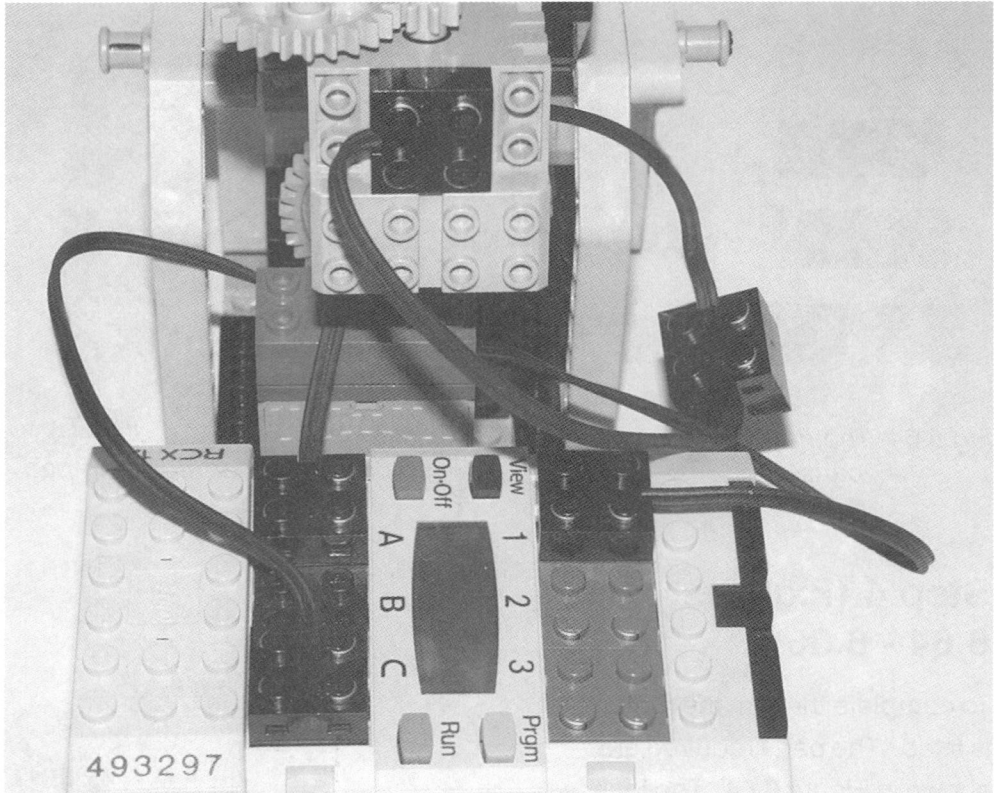

Fig.6.67 The connections to the RCX unit

connected to output C of the RCX unit, and the motor on the arm is connected to output A. A normal connecting lead is too short to make the connection to the motor on the arm. Either use one of the long connecting cables or join two standard leads to effectively produce a double length lead. The completed robot is shown in Figures 6.68 and 6.69.

Gripperbot software

Gripperbot will operate with the same software as Handybot, but the lack of slippage in the finger mechanism makes it necessary to reduce the operating time of the motor that drives the fingers. About 1.3 seconds seems to be sufficient, but it might be necessary to make some fine adjustment in order to

Fig.6.68 The completed Gripperbot with the fingers in the correct
           standby position

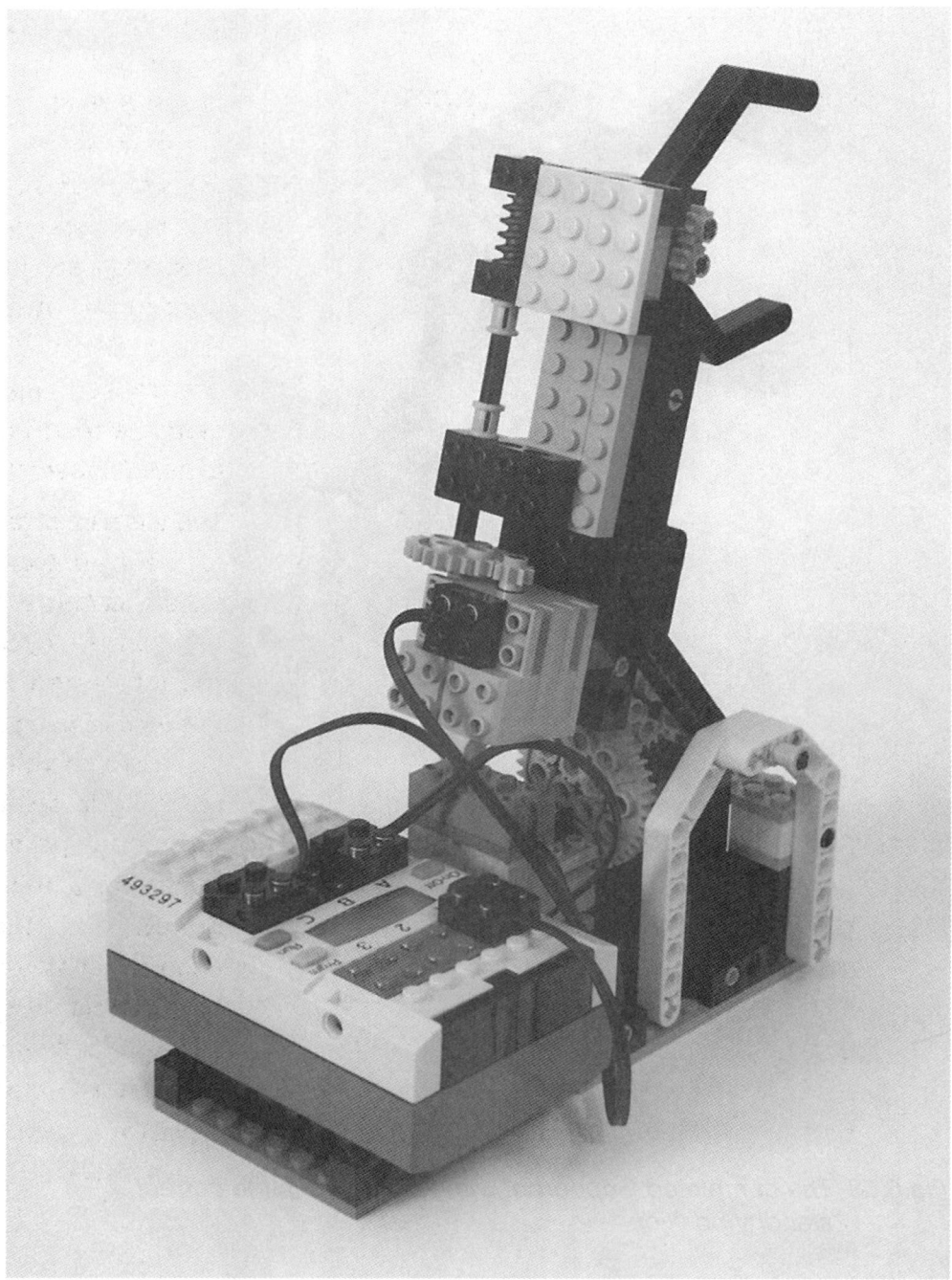

Fig.6.69 Another view of the completed Gripperbot robot arm

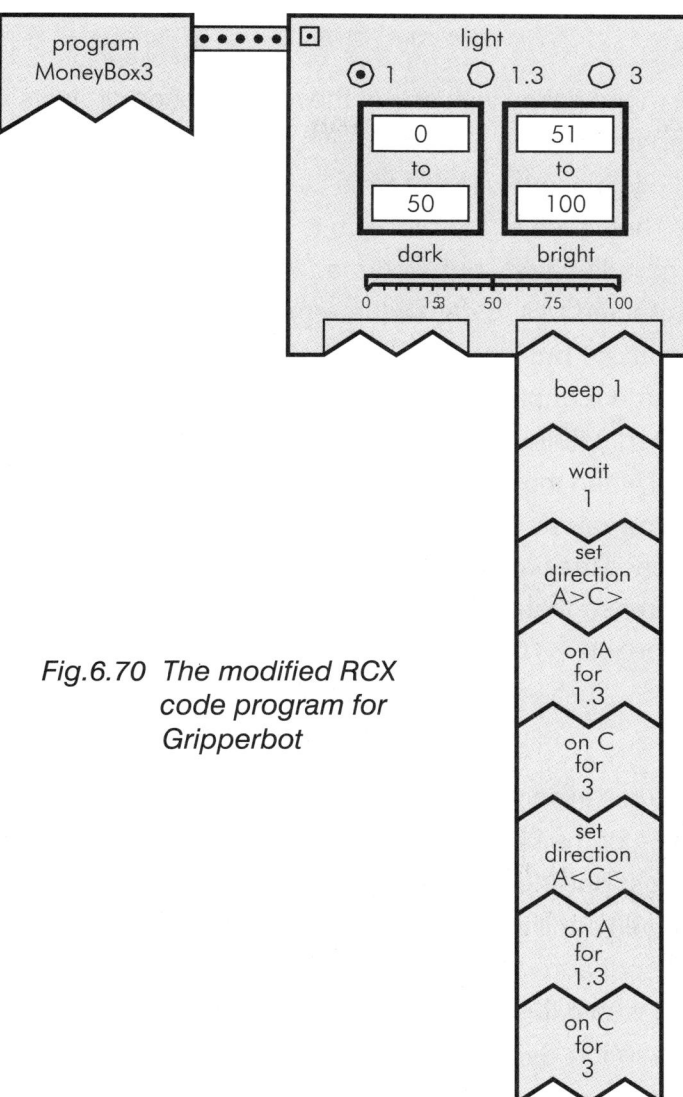

Fig.6.70 The modified RCX code program for Gripperbot

get the finger action just right. Figure 6.70 shows the modified RCX code for a version of Gripperbot that uses a light sensor. Clearly the other Handybot programs can easily be modified to suit Gripperbot.

There is a lot of fun to be had from experimenting with robot arms. If you are interested in this type of thing it is worthwhile investing in some additional parts, including a third motor. This enables the range of movements to be expanded, and the arm can then perform much more difficult tasks.

Problems

One slight problem you might notice with any of these arm-based projects is that after a number of operations the arm may not fully return to its standby position. This is most likely to occur with the arms that are fitted with a motor, since they are heavier. The cause of the problem is simply that gravity aids the motor when the arm is lowered, but works against it when the arm is raised. I would have expected this to cause the arm to significantly shift its position after each operation, but in practice the shift is minimal. This is presumably due to the large reduction ratio used in the drive mechanism for the arm. The sophisticated solution is to have a touch switch and a simple software routine to detect when the arm has reached its standby position, so that the arm will always be raised to exactly the same position. The more simple solution is to increase the time that the motor is switched on when it is raising the arm. Any increase that is required will be very small, and about 0.1 seconds should do the job.

Differential

The rover style robots featured in some of the previous chapters used individual drive motors for each side of the vehicle to permit steering. This is a rather crude way of doing things, but it does actually work very well. In fact this gives the ultimate in manoeuvrability. Some vehicles of this type can more or less spin around on the spot and do not have a turning circle in the normal sense of the term. If you would like to try something more conventional the parts in the Robotics Invention System are sufficient to build a "standard" vehicle having one motor to drive the rear wheels and another to provide steering. As we saw in chapter one, there are even some parts to build a differential so that the rear wheels can be driven in much the same way as they are on full-size "the real thing". This enables the outer rear wheel to rotate faster than the inner wheel when the vehicle turns. Rack and pinion steering is more difficult without the aid of some additional parts, but a steering system using a worm drive can be simple but effective. A simple worm drive steering system is used in Steeringbot, the robot featured here (Figure 6.71).

Fig.6.71 Steeringbot has front wheel steering provided by a worm drive

Step 1 (Figures 6.72 and 6.73)

Construction starts with the two sides of the chassis. Figure 6.72 shows one side that has been built, and one that is part assembled. Note that the two sections are "mirror images" of each other. The two 6 x 1 beams are pegged to

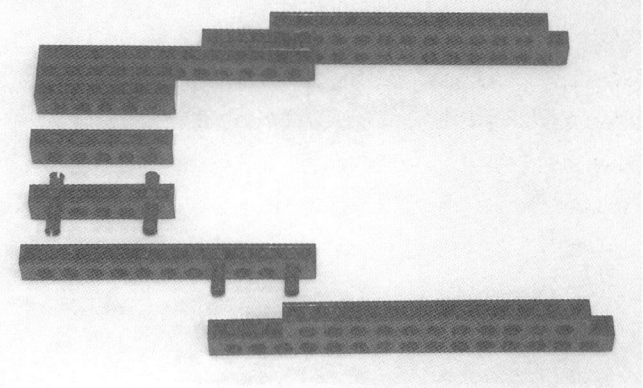

Fig.6.72 Completed and part assembled side sections

Fig.6.73 The sides joined together

each other and to one of the 12 x 1 beams using pegs that are three units long, and not the usual two unit long pegs. An 8 x 1 beam and a 10 x 1 beam are used to join the two sections together at the rear, as shown in Figure 6.73.

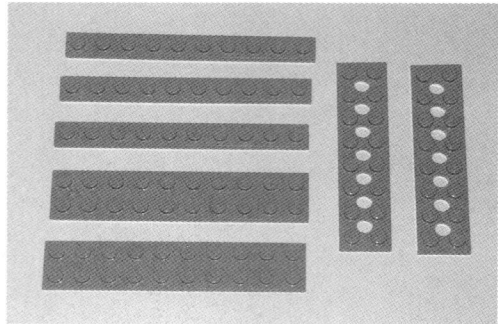

Fig.6.74 The plates required to strengthen the chassis

Step 2 (Figures 6.74 and 6.75)

The two sides of the chassis must now be more securely fixed together. The plates shown in Figure 6.74 are added onto the underside of the chassis to strengthen it. This gives the assembly shown in Figure 6.75.

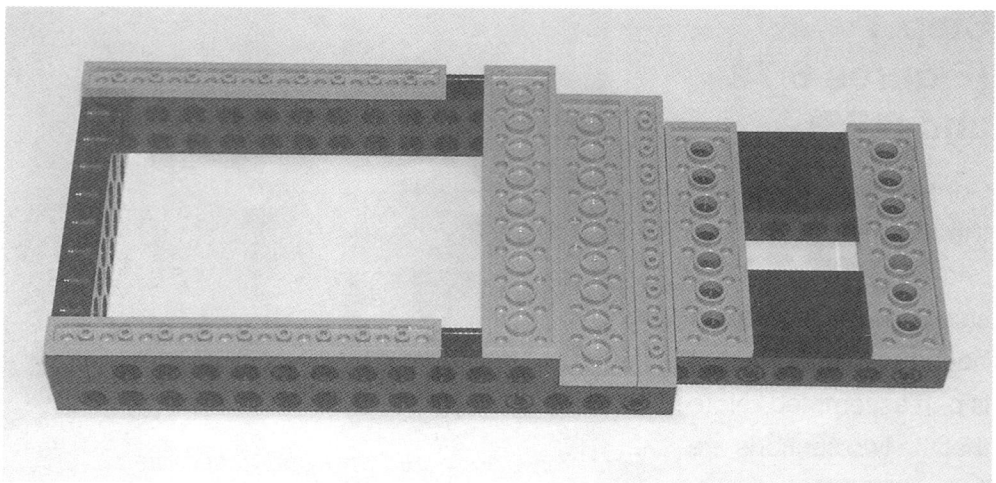

Fig.6.75 The plates fitted to the underside of the chassis

Step 3 (Figures 6.76 and 6.77)

Further plates and some beams (Figure 6.76) are added on the top of the chassis to help join the two sides together, and to form part of the platform that will eventually take the RCX unit. This gives the finished chassis of Figure 6.77.

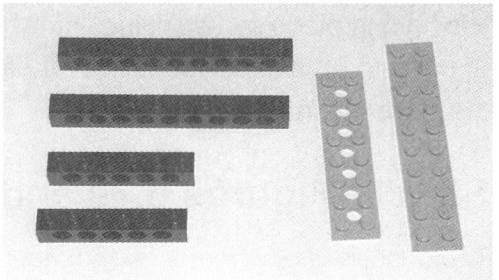

Fig.6.76 The plates and beams for the upper chassis

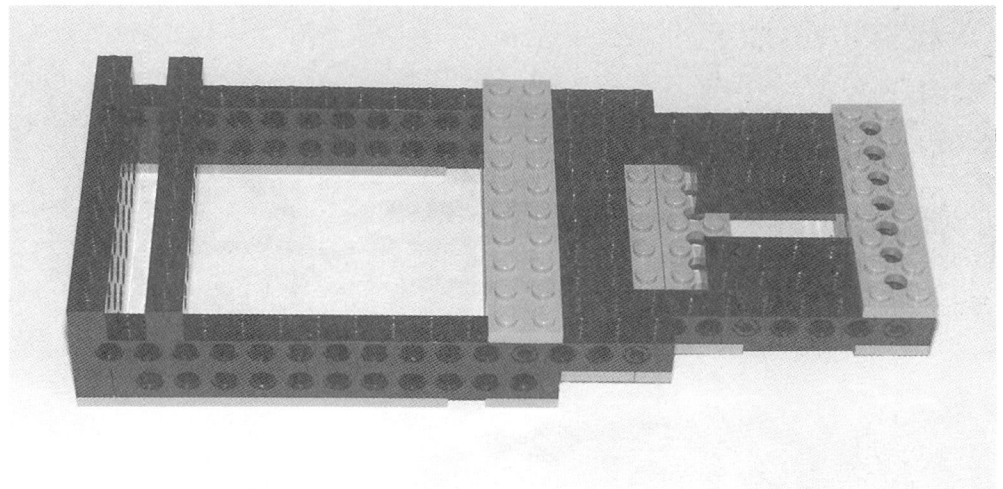

Fig.6.77 The completed chassis ready to take the wheels, drive shafts, etc.

Step 4 (Figure 6.78)

Next a drive shaft is fitted to the chassis. This must be fitted before the rear wheels are added, since the wheels block access to the holes into which the drive shaft is fitted. The shaft is 78 millimetres long, and

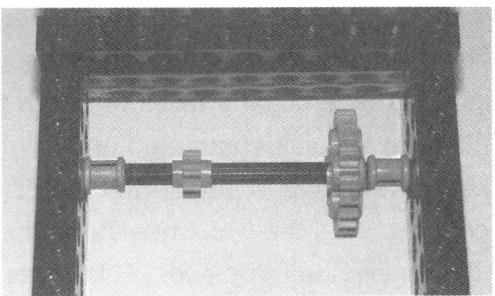

Fig.6.78 The drive shaft fitted to the chassis

it is held in place by two fixing "nuts" fitted inside the chassis. The positions of the 8-tooth and 12-tooth gearwheels can be "fine tuned" once the rest of the drive mechanism is in place.

Step 5 (Figures 6.79 and 6.80)

The chassis is now ready for the rear wheels and the differential to be added. This requires the components shown in Figure 6.79. Constructing the differential was covered in chapter 1, and it is admittedly a bit tricky. In fact it is trickier here because in must be constructed in situ. However, provided you remember to fit an angle gear onto the spigot in the differential before

Fig.6.79 The parts needed to build the differential and rear wheel assembly

fitting either of the axles and the other gears, everything should be reasonably straightforward. The axles are both 63 millimetres long incidentally. Once the differential is completed and in place, use a fixing "nut" on each axle to keep everything in place, with the differential positioned in the middle of the chassis. Add the two wheels and then two more fixing "nuts" to make sure the wheels stay in place. This gives the finished assembly of Figure 6.80. Make sure that the larger gear on the differential meshes properly with the

Fig.6.80 The differential and rear wheels fitted to the chassis

small gearwheel on the drive shaft. The larger gearwheel on the drive shaft must be pushed right to one side so that it does not come into contact with the differential.

Step 6 (Figures 6.81 - 6.83)

Now the drive motor is fitted in place and part of the mounting pad for the RCX unit is produced using the components shown in Figure 6.81. Start by adding three plates on the underside of the

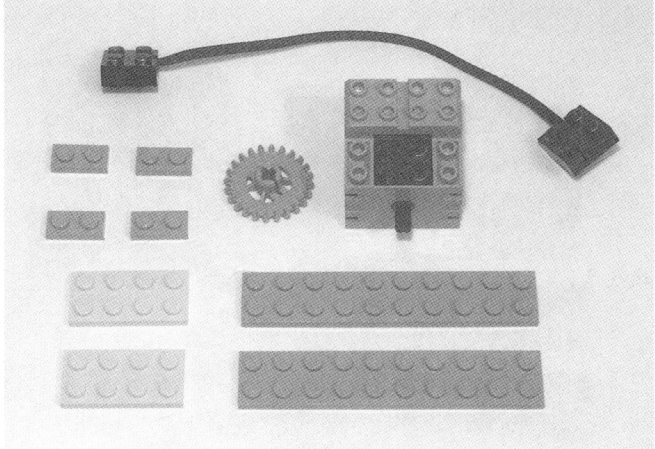

Fig.6.81 The parts for the motor assembly

chassis to make a mounting pad for the motor (Figure 6.82). The motor is fitted with a crown gear and is then installed on its mounting pad. Make sure the gears are aligned properly so that the crown gear meshes fully with the

Fig.6.82 The motor installed on
        its mounting pad

Fig.6.83 The finished assembly

existing 24-tooth gearwheel. Two 2 x 1 plates are now fitted on the top of the
chassis, one on each side, 8 and 9 units from the rear of the unit. Two more
2 x 1 plates are then added on top of these, and a lead is connected to the

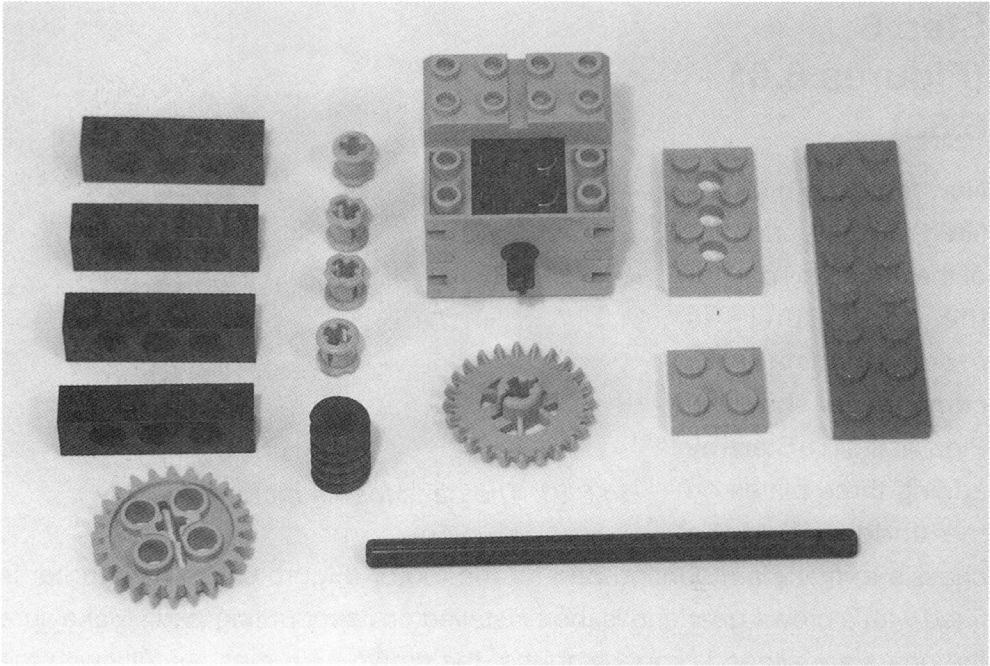

Fig.6.84 The steering motor and associated components

motor. A 10 x 2 plate is then added across the top of the 2 x 1 plates and the connector block on the motor, so that the motor is trapped in place from above and below (Figure 6.83).

Step 7 (Figures 6.84 - 6.86)

Next the drive for the steering is added, and this needs to parts shown in Figure 6.84. Start by fitting the three plates onto the chassis to form a mounting pad for the motor (Figure 6.85). Fit a crown gear on the motor and then install the motor on its mounting pad. Next fit two pairs of 4 x 1 beams to the chassis to act as supports for the worm gear. A 79-millimetre axle is then threaded through the central hole in one of the upper beams, and then it is fitted with a 24-tooth gearwheel, a fixing "nut", a worm gear, and another fixing "nut". With the axle fitted into the support on the other side of the chassis, two further fixing nuts are used to keep everything in place (Figure 6.86).

Fig.6.85 *The motor's mounting pad*

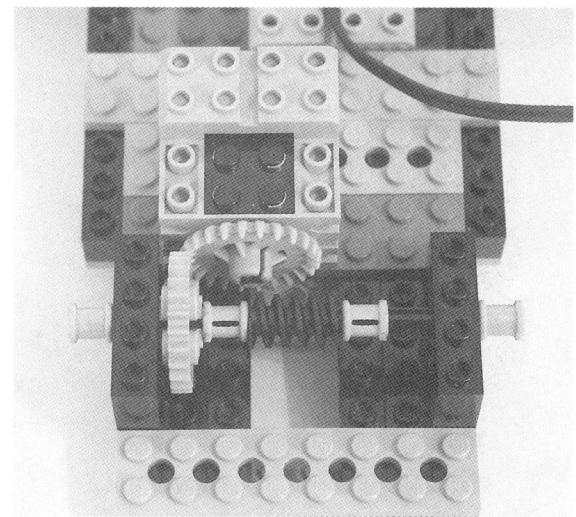

Fig.6.86 *The installed worm drive*

Fig.6.87 The parts for the front wheel assembly

Fig.6.88 The finished steering system

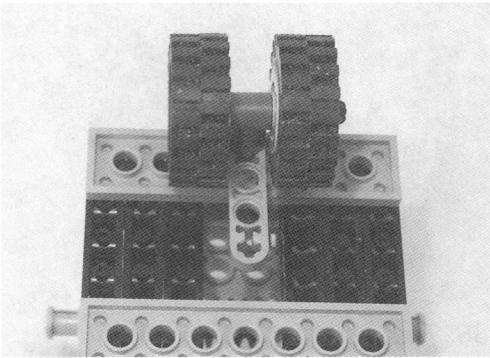

Fig.6.89 The front wheel assembly

Step 8 (Figures 6.87 - 6.89)

The robot is now ready for the front wheels to be added. Like some of the previous robot vehicles, the front wheels are mounted close together so that they effectively form one wide wheel. The parts required are shown in Figure 6.87, with the top and bottom sides of the finished assembly shown in Figures 6.88 and 6.89. The 31-millimetre axle is fitted through the blue plastic piece, and a 30-millimetre diameter wheel is then fitted at each end. Small pulleys can be added at the ends of the axle to act as fixing "nuts", but this is not essential. Make sure that this assembly is not too tight, and that the wheels can turn freely. The 46-millimetre axle is pushed into the end of the blue plastic piece, and then the grey girder piece is fitted onto the axle. It is pushed right down against the blue plastic piece, with the protrusion facing upwards. The axle is now pushed through the middle hole in the front section of the chassis, and a small pulley is added on the top side to lock the assembly in place. The 40-tooth gearwheel is then added at the top of the axle, and it is pushed down far enough to get it to mesh with the worm gear properly. A worm gear is well suited to this application since it locks the steering in the set position, and the high reduction ratio of 40 to 1 obtained here gives well controlled steering.

Step 9 (Figures 6.90 and 6.91)

The optional touch sensor is activated when the steering is at its central position. This is not essential when controlling the robot in real-time from a PC, but it is more than a little useful when a program running in the RCX unit controls the robot. Ideally a rotation sensor would monitor the steering so that the control software would always know the exact direction of the steering. The Lego MindStorms system does include a rotation sensor, but it is not included in the Robotics Invention System. The touch switch enables the

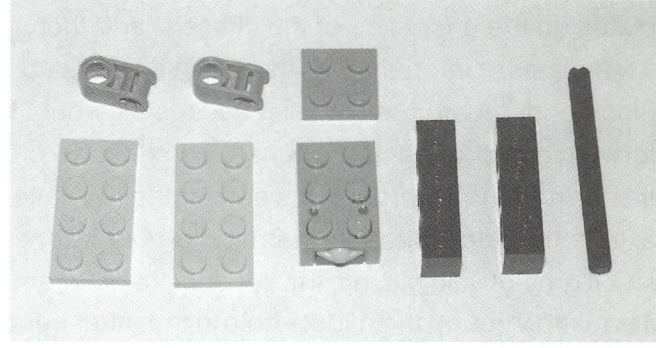

Fig.6.90 *The parts for the steering position sensor*

Fig.6.91 *The completed steering position sensor*

steering to be accurately returned to a central setting after a turn has been made, and this avoids having the steering gradually slip one way or the other until the robot is more or less going round in circles.

The parts needed to implement the touch sensor are shown in Figure 6.90, and Figure 6.91 shows the finished construction. Start by fitting the three plates on the underside of the chassis, and then fit the 4 x 1 beams onto the 4 x 2 plates. Unfortunately, the touch sensor can not simply be pressed into place underneath the chassis, because it would be either half a unit too far forward or too far back. Fortunately, Lego have provided a "get out clause" in the form of a hole through the sensor that will take an axle. The 46-millimetre axle is threaded through the middle hole in one of the beams, one of the small grey plastic pieces, the sensor, the other small plastic piece, and finally it is taken through the middle hole in the other beam. The small plastic pieces prevent the sensor from flipping upward out of position.

Fig.6.92 The RCX unit "looks" backwards

Step 10 (Figure 6.92)

The RCX unit is now fitted at the rear of the chassis. I suggest having the windows for the infrared communicator facing backwards where they will be totally free of any obstructions.

Step 11 (Figures 6.93 - 95)

The optional light sensor permits Steeringbot to be used as a line following robot. It is fixed to the chassis using a 2 x 4 right-angled plate (Figure 6.93). The parts required are shown in Figure 6.94.

Fig.6.93 *The sensor on its angle plate*

The right angled plate fits on the underside of the chassis, on the left-hand side, four and five units back from the front of the unit. The 2 x 1 plate fits immediately in front of it. The 10 x 1 plate is then added, and it serves to trap the right angled plate firmly in place. The light sensor is then fitted onto the right-angled bracket, and it is connected to input 3 of the RCX unit. To complete the

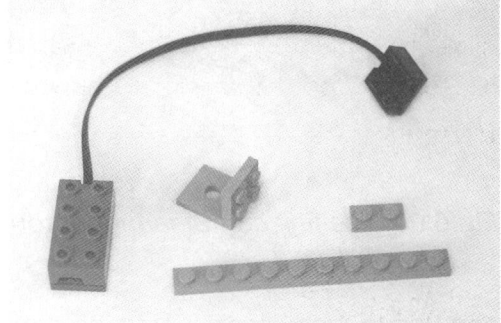

Fig.6.94 *The parts for the sensor assembly*

unit connect the drive motor to output A, the steering motor to output C, and the touch sensor (if fitted) to input 1 (Figure 6.95). A standard connecting lead is not long enough to make the connection from the touch sensor to the RCX unit. One of the long connecting leads can be used, but it is more practical to connect two of the short leads together. The two blocks at the point where the cables are joined can be

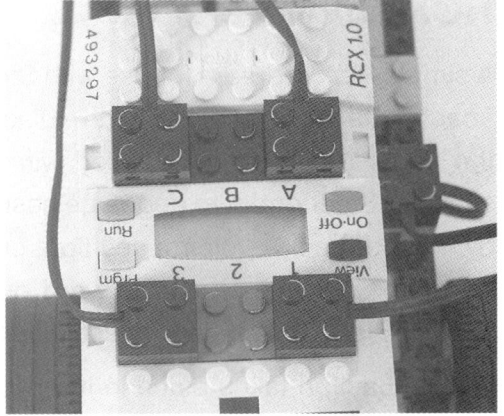

Fig.6.95 *Connections to the RCX unit*

Fig.6.96 The finished Steeringbot robot

secured to the chassis just to the right of the RCX unit. The completed robot is shown in Figures 6.96 and 6.97.

RCX software

A simple RCX program for use with Steeringbot is shown in Figure 6.98. This enables the robot to follow inside a black line in a clockwise direction. However, the paper track that is supplied with the Robotics Invention System is too small for Steeringbot to negotiate easily, so I would recommend making your own larger track, avoiding any tight curves. The main program simply turns on motor A and sets it to forward operation. This sets Steeringbot going forwards, and the steering must initially be set at a central setting.

The real action takes place in the right-hand section of the program, in the blocks headed by the light sensor watcher. This waits for a reading of 40 or less from the light sensor, but as always, you may have to adjust this figure to

Fig.6.97 Rear view of the finished robot

suit the particular light sensor and track you are using. When the light sensor goes over the black line the reading will go below the threshold level and the series of instructions will then be performed. First motor A is switched off to stop the robot and prevent it from running over the line. Bear in mind that, unlike the robots that steer using the driving wheels, it takes Steeringbot a while to get the steering well over to one side so that it can turn sharply.

Next the direction of motor C is set to reverse, which in this context actually means the steering is adjusted to the right. Motor C is then switched on for 0.7 seconds to actually set the steering well to the right. The time used here might need some fine adjustment to produce optimum results. Motor A is then switched on again to set the robot moving forwards and to the right. The direction of motor C is set to forward operation and it is switched on, which

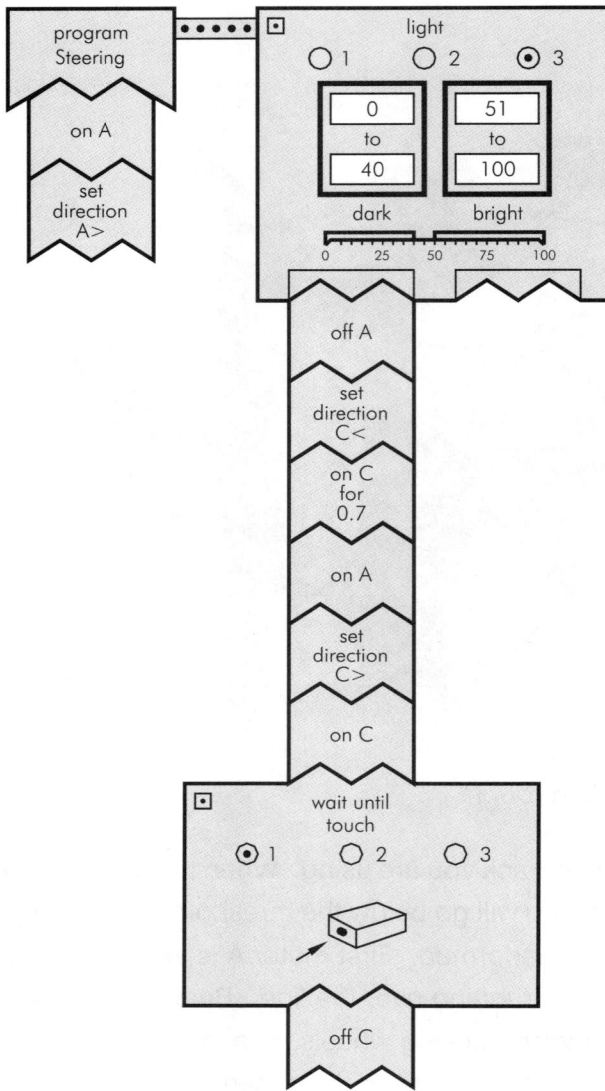

Fig.6.98 *The Steeringbot RCX code*

starts to adjust the steering back towards the central setting. This continues until the "wait until" block detects that the touch switch has been operated, which means that the steering is back at its central setting. Motor C is then switched off to hold the steering at this central setting, and the robot continues forward until the light sensor watcher detects the black line again. The whole process is then repeated.

A robot such as Steeringbot is never going to be as manoeuvrable as one which has steering provided by the drive wheels. Nevertheless, it could presumably be made to negotiate a tighter track with some changes to the software. One possibility is to actually reverse the robot rather than just halting it when the line is detected. This would give it more room to manoeuvre and should enable it to tackle tighter curves. The price that would be paid for the room to manoeuvre is a greatly reduced rate of progress around the track.

VB Software

This program in Visual BASIC provides a simple remote control for Steeringbot:

```
Private Sub Command1_Click()
Spirit1.SetFwd "2"
Spirit1.On "2"
End Sub

Private Sub Command2_Click()
Spirit1.SetRwd "2"
Spirit1.On "2"
End Sub

Private Sub Command3_Click()
Spirit1.Off "0"
End Sub

Private Sub Command4_Click()
Spirit1.On "0"
End Sub

Private Sub Command5_Click()
Spirit1.SetFwd "0"
End Sub

Private Sub Command6_Click()
Spirit1.SetRwd "0"
End Sub

Private Sub Command7_Click()
Spirit1.CloseComm
End
End Sub
```

```
Private Sub Command8_Click()
Spirit1.Off "2"
End Sub

Private Sub Form_Load()
Spirit1.InitComm
End Sub
```

This program requires eight command buttons, as can be seen from Figure 6.99, which shows the program in operation. The table provided here shows the correlation between each button's name and its caption:

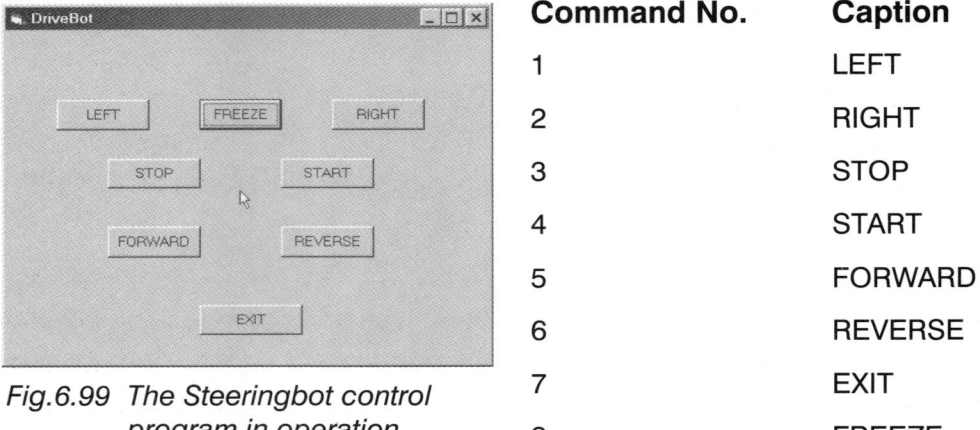

Command No.	Caption
1	LEFT
2	RIGHT
3	STOP
4	START
5	FORWARD
6	REVERSE
7	EXIT
8	FREEZE

Fig.6.99 The Steeringbot control program in operation

Operating most buttons sends the appropriate On or Off instruction to the RCX unit, complete with a direction command where appropriate. The FREEZE button switches off the steering motor to freeze the steering at its current position. The EXIT button, as usual, closes communications with the serial port and closes the program.

Walk it

Robots that walk are guaranteed to raise a smile from your family and friends, but in terms of the technical challenges they provide, walking robots are several orders more difficult than wheeled or tracked vehicles. In theory a vehicle that walks is capable of traversing rough terrain that wheeled or even tracked vehicles would find practically impossible. Animals can freely wander over difficult terrain that would soon defeat most wheeled or tracked vehicles. However, animals are complex machines with very elaborate control systems. Emulating this type of thing with a robot is extremely difficult, but it is perhaps the control mechanism that provides the greatest challenge. Making a robot that can walk is one thing, but giving it something analogous to a sense of balance is quite another. Many walking robots seem quite plausible until they reach the smallest of obstacles and promptly fall over. In fact some fall over when walking over perfectly flat ground!

A complex walking robot probably requires something more than an unexpanded Robotics Invention System, but with the basic kit you can certainly make a start with this type of thing. The simplest type of walking robot uses the system outlined in Figure 6.100. It consists of a chassis fitted with two wheels on each side, or cams will do the job just as well. The wheels on each side are fitted with two legs in a single piece. The legs are pivoted on the wheels but the pivot points are off-centre.

Figure 6.100 shows a side view, so only two wheels and one leg section can be seen. In the top diagram the legs are in their highest position with the pivot points at the top, and the legs are held clear of the ground. In the next diagram the wheels have moved on by 90 degrees in a clockwise direction so that the legs have moved forwards and downwards and now just touch the ground. The legs have moved forwards, but the chassis remains unmoved. Things are different at the next stage where the legs are in the same position, but the chassis has been lifted upwards and forwards. In the final stage things have moved on by a further 90 degrees, which has taken the chassis further forward but downwards, so that it is now back down on the ground. The important point to note is that the chassis, and therefore the entire robot, has moved forward. On each turn of the wheels the robot lifts itself up and down, and in the process carries itself forward slightly. Of course, simply reversing the direction of the wheels will move it in the opposite direction.

A robot of this type is fun to try out, but it has major limitations. Something you tend to notice when dealing with walking robots in general is that they are much slower than wheeled or even tracked counterparts. In fact they tend to be painfully slow. In this case it is not difficult to work out the reason for the lack of speed. The robot has to lift virtually its whole weight into the air each time it moves forward. This wastes an enormous amount of energy. Any form of stop-go movement tends to be inefficient anyway, with the vehicle not building

Fig.6.100 This probably represents the most simple form of walking mechanism

up any momentum. You could speed things up by using sheer brute force, but this does not usually work in practice. Running robots have to be very well engineered or they simply self-destruct.

Walkbot

Walkbot (Figure 6.101) is based on the principle outlined in Figure 6.100. It works quite well, if rather slowly, but it only has forward and reverse movement. Steering with this method of propulsion is difficult rather than impossible. The obvious way of implementing steering is to use separate motors and drive systems for the two sides. With the legs in the up position on one side,

Fig.6.101 Walkbot moves around on its four legs, but very slowly

operating the legs on the other side should gradually shuffle the robot round to the required direction. Do not overlook the fact that the legs only work well when the two sides are properly synchronised. It would be easy to make a robot that could steer but tended to flail around and "drown" rather than move backward or forward. Probably the best way to steer is to have a rotating base on the chassis. With both sets of legs off the ground the base would enable the robot to be swung round the face in the right direction. It could then move off in that direction by operating the legs in the normal way. This system gives excellent manoeuvrability, but could easily make the robot so heavy that it would move very slowly indeed. Anyway, Walkbot represents an amusing project as it stands (on its own four feet), and an interesting starting point if you wish to experiment with walking robots.

Fig.6.102 Each side consists of three beams

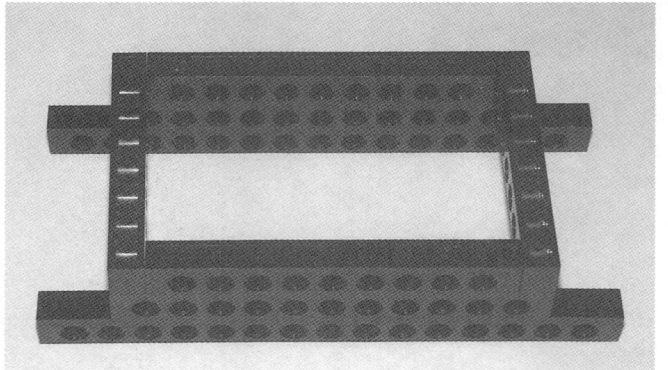

Fig.6.103 The completed basic chassis

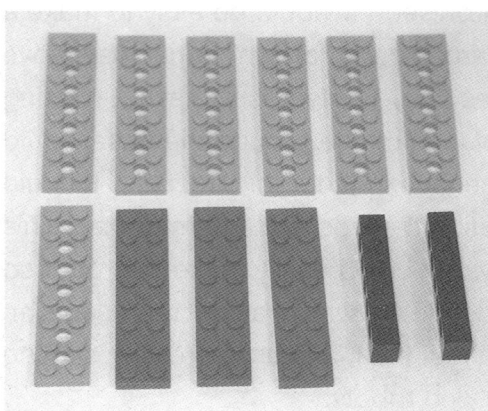

Fig.6.104 The chassis plates

Step 1 (Figures 6.102 and 6.103)

As usual, the first step is to construct a basic chassis. This has side pieces that are each constructed from three beams. Figure 6.102 shows the parts for one side plus a completed side section. The two side sections are joined by two 8 x 1 beams added on the underside, as shown in Figure 6.103.

Step 2 (Figures 6.104 - 6.106)

The basic chassis now needs to be bolstered slightly by the addition of two beams and numerous plates (Figure 6.104). A 6 x 1 beam is added at each end of the chassis, with an 8 x 2 plate above and below each one to sandwich it in place.

Figure 6.105 shows one of these beams in position, complete with the two plates. Six 8 x 2 plates are then fitted on the underside of the chassis (Figure 6.106).

Fig.6.105 One of the end beams

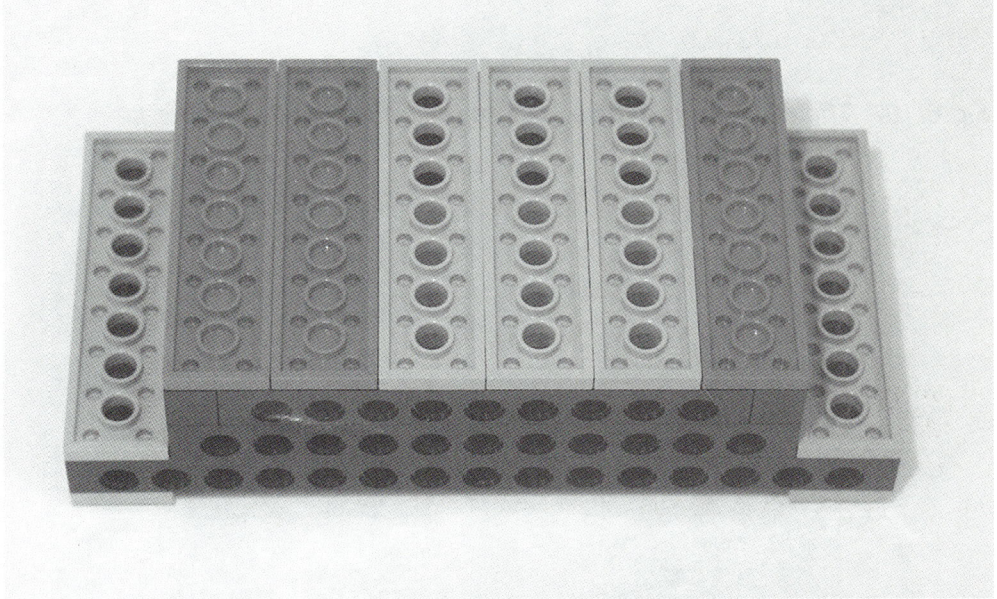

Fig.6.106 The completed chassis is simple but tough

Step 3 (Figures 6.107 and 6.108)

Three shafts complete with various gears are now added to the chassis. Figure 6.107 shows the parts required and the complete assembly is shown in Figure 6.108. The shorter shaft is 79 millimetres long and the longer ones are 95 millimetres in length. The 40-tooth gearwheels act as the cams that drive the

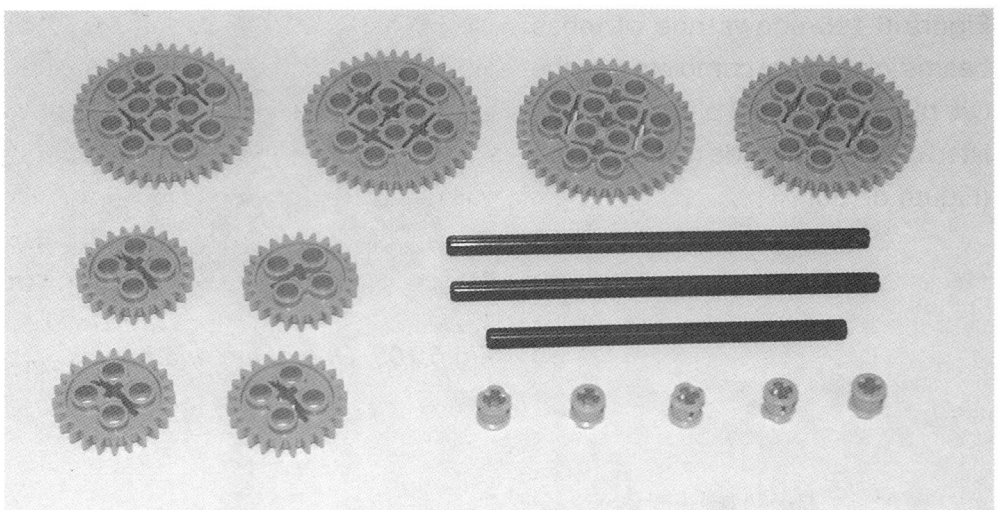

Fig.6.107 *The parts for the drive system that operates the legs*

Fig.6.108 *The finished drive system in position*

legs. The central shaft and gear is needed to link the other two shafts. This method of propulsion requires all the cams to be driven, so this link is essential. You can not simply link the two outer shafts directly with gears, as they would then turn in opposite directions. The 24-tooth gearwheel that serves no apparent purpose at present will eventually take the drive from the motor via a worm gear. The 24 to 1 reduction from the motor ensures that Walkbot has sufficient oomph to move itself, but do not expect it to go very fast. Fitting the axles and gears is easy enough, but there is a trap that must be avoided. Make sure that all four of the 40-tooth gearwheels have precisely the same orientation, or it will not be possible to fit the leg assemblies later on.

Step 4 (Figures 6.109 and 6.110)

Some plates must be added to the chassis to form the basis of the drive mechanism. The plates required are shown in Figure 6.109 and the finished assembly can be seen in Figure 6.110.

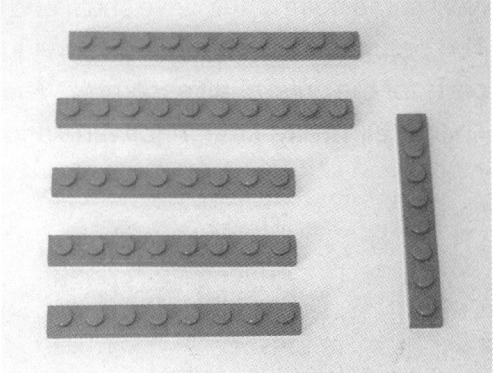

Fig.6.109 The plates for the drive mechanism

Fig.6.110 The plates are added to the chassis where they form the basis for the drive mechanism

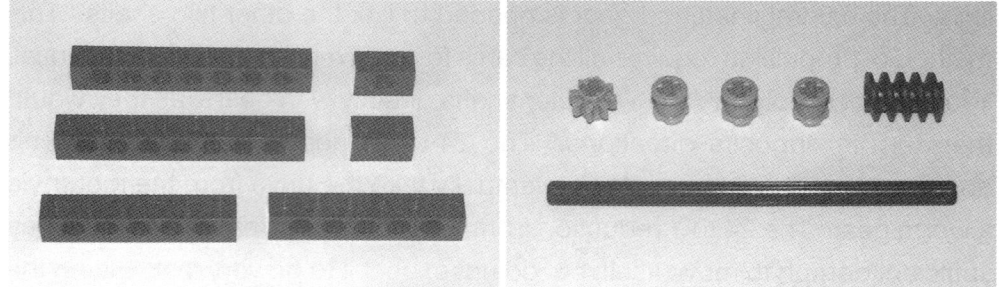

Fig.6.111 The support beams   *Fig.6.112 The parts for the worm drive*

Step 5 (Figures 6.111 - 6.113)

Next some beams, the drive shaft, and the worm drive are added to the chassis. The beams required are shown in Figure 6.111, and Figure 6.112 shows the parts for the drive shaft assembly. The shaft is 79 millimetres long incidentally. This is all pretty straightforward and merits little expansion. Figure 6.113

Fig.6.113 The worm drive is complete, but requires strengthening

shows the chassis with the beams and drive assembly added. The position of the 24-tooth gearwheel might need slight adjustment to get it to mesh properly with the worm gear.

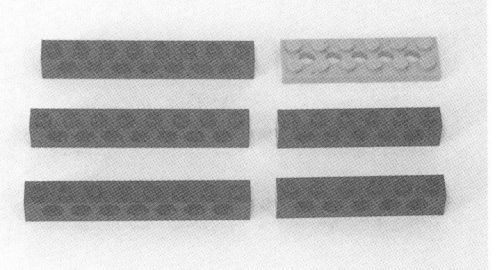

Fig.6.114 *The parts to reinforce the worm drive*

Step 6 (Figures 6.114 - 6.116)

As things stand, the worm drive is fitted and working, but this part of the unit is rather flimsy. Five beams and a plate (Figure 6.114) are now added to give firmer support for the worm drive, and to provide the rear platform to support the RCX unit. Start by fitting x 6 x 1 beam at the side of the chassis, and a 6 x 1 beam plus an 8 x 1 type across the chassis (Figure 6.115). The last two partially fit on top of the two 6 x 1 beams that go across the chassis and form part of the support frame for the worm drive. Make sure all the beams fit together properly and that none of the existing components are knocked out of position. Two 8 x 1 beams are then added across chassis, and the 6 x 2 plate is added on top, giving the finished assembly of Figure 6.116.

Fig.6.115 *The initial reinforcement in place*

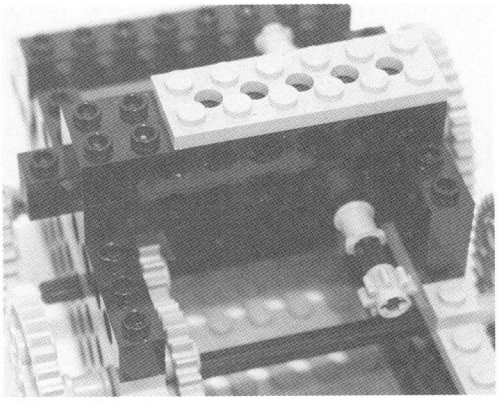

Fig.6.116 *The completed worm drive*

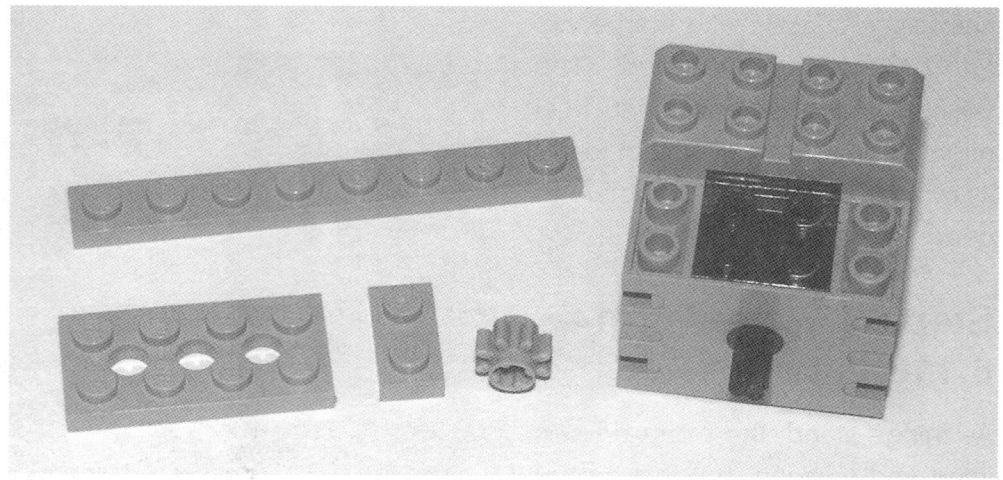

Fig.6.117 The motor and parts for its mounting pad

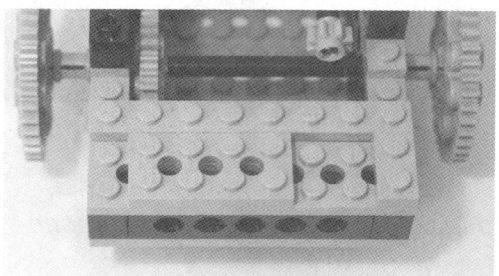

Fig.6.118 The pad in position

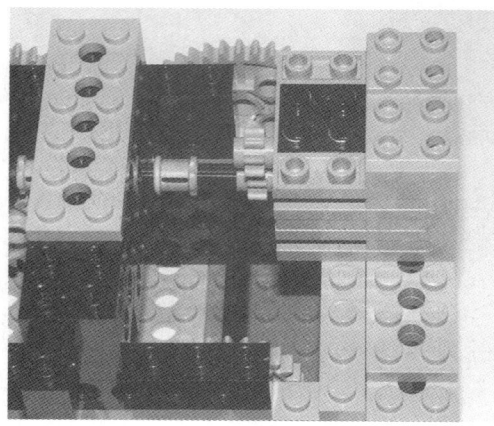

Step 7 (Figures 6.117 - 6.119)

The motor is fitted onto a mounting pad at the rear of the chassis. Figure 6.117 shows the parts required and Figure 6.118 shows the completed mounting pad. An 8-tooth gearwheel is fitted onto the motor's shaft and the motor is then fitted onto its mounting pad, making sure that the two 8-tooth gears mesh together properly (Figure 6.119).

Fig.6.119 The motor fitted on its mounting pad and coupled to the worm drive

Step 8 (Figures 6.120 - 6.122)

Next the light sensor and the RCX unit are fitted. The light sensor forms part of the mounting for the RCX unit, so it must be included even if its services as a sensor are not required. It could probably be replaced with two 2 x 2 beams and a 4 x 2 plate, but I have not tried this. The parts required for the sensor assembly are shown in Figure 6.120, and the completed assembly in position on the chassis appears in Figure 6.121. The 4 x 2 plate is fitted centrally on the plate already present on the front of the chassis, and the 2 x 2 block is placed on top of this. Next the 8 x 2 plate is fitted on the beam near the front of the chassis, leaving the rear of the plate unsupported. The delta shaped plate is then fitted in front of this on the 2 x 2 block, and the light sensor is added onto these two plates. The front of the sensor should be flush with the front of the delta shaped plate. The two remaining 4 x 2 plates are then fitted on top of the light sensor. Next the RCX unit is added on top of the light sensor assembly and the 6 x 2 plate near the middle of the chassis (Figure 6.122).

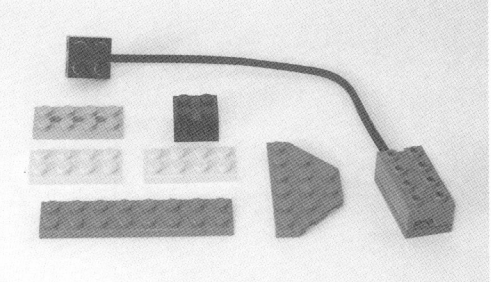

Fig.6.120 The light sensor and its associated parts

Fig.6.121 The sensor assembly in position

Fig.6.122 The RCX unit in position

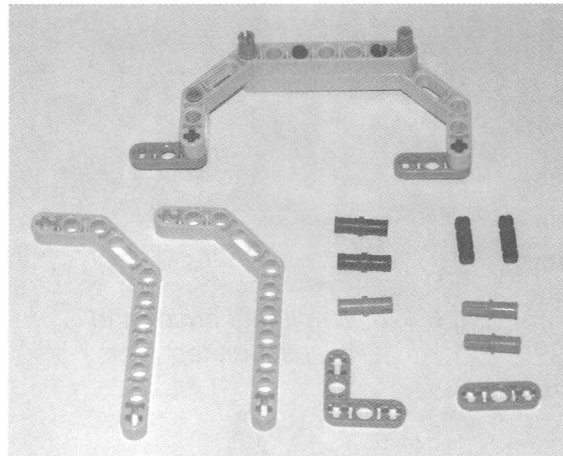

Fig.6.123 A leg assembly and the parts for one

Fig.6.124 An installed leg assembly

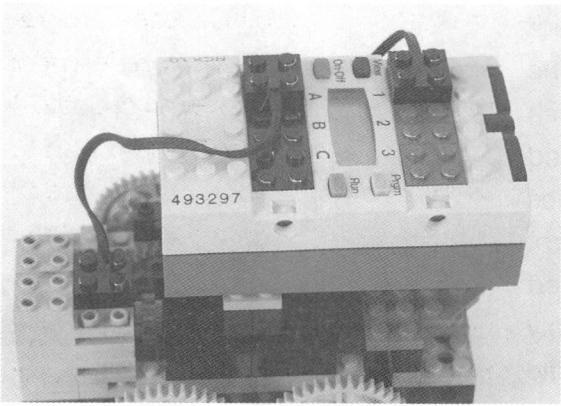

Step 9 (Figures 6.123 - 6.125)

The legs and feet are now added to the large gearwheels. The parts required for one leg assembly are shown in Figure 6.123, together with one completed assembly. This is all fairly straightforward, but note that several of the pegs used are not the standard variety. Some are like short axles, and others have the standard fitting at one end and an axle style fitting at the other. Figure 6.124 shows a completed leg assembly fitted to the robot, and the other leg assembly is much the same. To complete Walkbot connect the motor to output A of the RCX unit, and the light sensor to input 1 (Figure 6.125). Two views of the finished robot appear in Figures 6.126 and 6.127.

Fig.6.125 The connections to the RCX unit

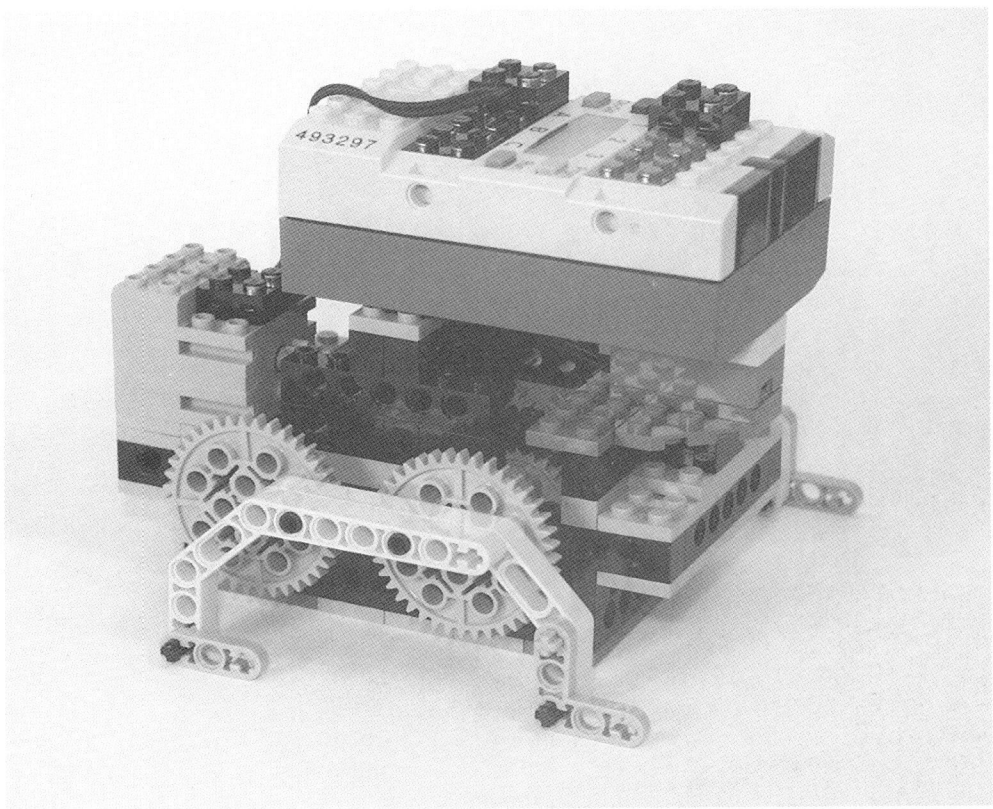

Fig.6.126 The finished Walkbot, raring to go

Walkbot RCX code

As Walkbot can only go backwards and forwards it does not require particularly sophisticated control software. The main section in the RCX code program of Figure 6.128 starts by switching on the motor, and setting it to go forwards. Next the timer is reset, and then the program waits for the timer to count to 40 seconds, after which the motor is switched off. This is either to stop Walkbot from bounding off and getting away, or it is just an opportunity to use the timer function. When you try out Walkbot you will soon discover which of these is the true reason.

The sensor watcher checks for a light level of 51 or more. If the robot gets close to an object it is likely that the light reflected from the object will take the

Fig.6.127 A rear view of Walkbot

reading above the threshold level. A "beep" sound is then emitted and the motor is switched off so that a collision is avoided. As usual, you may need to adjust the light threshold value to get things working properly.

Walkbot VB software

The Visual BASIC program for Walkbot makes cleverer use of the timers. Walkbot is set going forwards, and it continues to do so until the light sensor detects that it is approaching an object. The motor is then set into reverse, and Walkbot returns to its starting point. A timer is used to measure the time that the robot went forwards, and is then used to time the return journey so that it is the same length. This is the Visual BASIC listing for Walkbot:

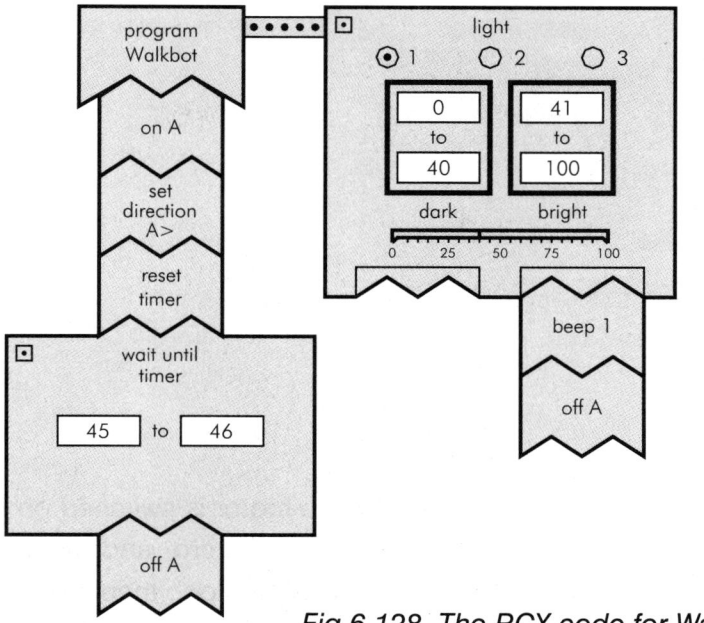

Fig.6.128 The RCX code for Walkbot

```
Private Sub Command1_Click()
With Spirit1
.InitComm
.SelectPrgm 1
.BeginOfTask 0
.SetSensorType 0, 3
.SetSensorMode 0, 4, 0
.On "0"
.SetFwd "0"
.ClearTimer 0
.While 9, 0, 1, 2, 50
.EndWhile
.SetVar 0, 1, 0
.ClearTimer 0
.SetRwd "0"
.While 0, 0, 0, 1, 0
```

```
      .EndWhile
      .Off "0"
      .EndOfTask
      End With

      End Sub

      Private Sub Command2_Click()
      Spirit1.CloseComm
      End
      End Sub
```

After the usual setting up processes the motor is switched on and set to go forwards. Timer 0 is then cleared (reset to zero) and this command also restarts the timer. An empty While...EndWhile loop then repeatedly monitors the light sensor and prevents the program from progressing further until a light level of 50 or more is detected. Once again, this threshold level may have to be "fine tuned" in order to obtain good results. Once a suitable light level has been detected the program moves on, and variable 0 is set at the value read from Timer 0. The timer is then reset to zero, and the motor is set into reverse. Next another empty While...EndWhile loop repeatedly compares the value stored in variable 0 with the value read from Timer 0. It continues to loop while the value stored in variable 0 is greater than the one read from Timer 0, but eventually the value in Timer 0 will exceed the one in variable 0. The program then moves on, with the motor being switched off and the program terminating. Due to Walkbot's slow rate of progress there is little risk of overshoot occurring, and the robot should return quite accurately to its starting point.

Swaybot

If you would like to make Walkbot look more animal-like you could try the obvious ploy of having the two leg assemblies operate 180-degrees out-of-phase. In other words, fit the legs onto the 40-tooth gearwheels so that when one side is in the highest position the other is in the lowest (Figure 6.129). Unfortunately, this system does not work too well in practice because the chassis is not being properly lifted upward and forward. The real problem is that the chassis is never properly grounded. The robot just tends to just sway from side to side doing a dance

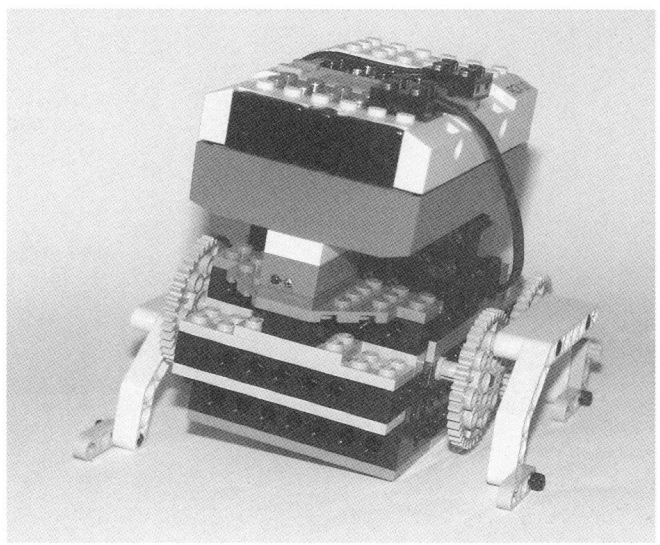

Fig.6.129 A more animal-like action, but little progress is made

rather than something approximating to a true walking action. If you try this system and the robot manages to make any progress, it is probably because it is on a slope! If you wish to try your hand at building more sophisticated walking robots you will probably have to use more than four legs. I believe that six legs is said to be the theoretical optimum, so Insectbot would seem to be the place to start.

Appendix 1

Web addresses

There are numerous web sites devoted to various aspects of Lego products, including some that are specifically devoted to Lego MindStorms robots and programming the robots in various languages. Any good search engine should soon turn up a large number of interesting sites. The few sites mentioned here are likely to be especially useful.

www.legomindstorms.com

This is the official Lego MindStorms site, and as one would expect, it has all the latest news, projects to build, etc.

www.legoworldshop.com

This is an online shop that sells Lego kits, spares, additional sensors and motors, etc. Those living in the USA and Canada should use the Lego Shop-at-Home Service, which is a telephone based mail order service. The telephone numbers are 1-800-835-4386 (USA) and 1-800-267-5346 (Canada).

www.geocities.com/area51//nebula8488/lego.html

This is the source of PBrickCommand, the program mentioned in chapter 2. Even if you have Visual BASIC or VBA it is well worthwhile giving this freeware program a tryout.

www.lego.com/dacta/robolab

Robolab™ is an educational version of MindStorms software and is supplied by Commotions Ltd, Unit 11, Tannery Road, Tonbridge, Kent, TN9 1RF, telephone 01732-773399. Commotions Ltd. are suppliers of educational robotics materials

Appendix 2

Useful facts and figures

Sources for commands such as SetVal, IF, and While.

Number	Source selected
0	Variable
1	Timer
2	Constant
3	Motor status
4	Random
8	Program number
9	Sensor value
10	Sensor type
11	Sensor mode
12	Sensor (raw)
13	Sensor (Boolean)
14	Watch
15	PB Message

Relational operators in If and While commands.

Number	Operator
0	> (greater than)
1	< (less than)
2	= (equal to)
3	<> (not equal to)

Variables are numbered from 0 to 31 and can contain values from −32768 to +32767.

Constants can also contain values from −32768 to +32767.

There are four timers numbered from 0 to 3.

Sensor numbers from 0 to 2 refer to inputs 1 to 3 respectively of the RCX unit.

Output numbers from 0 to 2 refer to outputs 1 to 3 on the RCX unit.

Appendix 3

Immediate commands

AbsVar

AlterDir

AndVar

BeginOfSub

BeginOfTask

ClearEvent

ClearSensorValue

ClearTimer

CloseComm

DatologNext

DeleteAllSubs

DeleteAllTasks

DeleteSub

DeleteTask

DivVar

DownloadFirmware

EndOfSub

EndOfSubNoDownload

EndOfTask

EndOfTaskNoDownload

Float

GetThreadPriority

InitComm

MemMap

MulVar

Off

On

OrVar

PBAliveOrNot

PBBattery

PBPowerDownTime

PBTurnOff

PBTxPower

PlaySystemSound

PlayTone

Poll

SelectDisplay

SelectPrgm

SetDatalog

SetEvent

SetFwd

SetPower

SetRwd

SetSensorMode

SetSensorType

SetThreadPriority

SetVar

SetWatch

SgnVar

StartTask

StopAllTasks

StopTask

SubVar

SumVar

TowerAlive

TowerAndCableConnected

UnlockFirmware

UnlockPBrick

UploadDatalog

Appendix 4

Downloadable commands

AbsVar

AlterDir

AndVar

ClearPBMessage

ClearSensorValue

ClearTimer

DatologNext

DivVar

Else

EndIf

EndLoop

EndWhile

Float

Gosub

If

Loop

MulVar

Off

On

OrVar

PBTurnOff

PBTxPower

PlaySystemSound

PlayTone

SelectDisplay

SendPBMessage

SetFwd

SetPower

SetRwd

SetSensorMode

SetSensorType

SetVar

SetWatch

SgnVar

StartTask

StopAllTasks

StopTask

SubVar

SumVar

Wait

While

Appendix 5

Sensor modes and types

The SetSensorMode mode settings.

Number	Mode selected
0	Raw
1	Boolean
2	Transition counter (counts changes from 0 to 1 and 1 to 0)
3	Periodic counter (counts pulses)
4	Percent
5	Celsius
6	Fahrenheit
7	Angle

The SetSensorType type settings

Number	Type selected
0	None
1	Switch (touch sensor)
2	Temperature
3	Reflection (light sensor)
4	Angle (rotation sensor)

Index